Advances on Modeling in Tissue Engineering

T0181142

Computational Methods in Applied Sciences

Volume 20

Series Editor

E. Oñate
International Center for Numerical Methods in Engineering (CIMNE)
Technical University of Catalonia (UPC)
Edificio C-1, Campus Norte UPC
Gran Capitán, s/n
08034 Barcelona, Spain
onate@cimne.upc.edu
www.cimne.com

For other titles published in this series, go to
www.springer.com/series/6899

Paulo R. Fernandes • Paulo Jorge Bártolo
Editors

Advances on Modeling in Tissue Engineering

 Springer

Editors
Paulo R. Fernandes
Instituto Superior Técnico
Technical University of Lisbon
IDMEC – Instituto de Engenharia
Mecânica
Av. Rovisco Pais 1
1049-001 Lisboa
Portugal
prfernan@dem.ist.utl.pt

Paulo Jorge Bártolo
Polytechnic Institute of Leiria
Centre for Rapid and Sustainable Product
Leiria
Portugal
pbartolo@estg.iplei.pt

ISSN 1871-3033
ISBN 978-94-007-3621-4   ISBN 978-94-007-1254-6 (eBook)
DOI 10.1007/978-94-007-1254-6
Springer Dordrecht Heidelberg London New York

Cover design: SPi Publisher Services

Printed on acid-free paper

Springer is part of Springer Science+Business Media (www.springer.com)

Contents

Preface   vii

Cell Mechanics: The Role of Simulation   1
Christopher R. Jacobs and Daniel J. Kelly

**Multiscale Modelling of Bone Tissue – Remodelling
and Application to Scaffold Design**   15
Helder C. Rodrigues, Pedro G. Coelho, Paulo R. Fernandes

**Nonlinear Elastic Scaffold Design, Modeling and Fabrication
for Soft Tissue Engineering**   35
Scott J. Hollister, Claire G. Jeong, J. Jakob Schwiedrzik,
Anna G. Mitsak, Heesuk Kang, Francesco Migneco

**Computational Techniques for Selection of Biomaterial Scaffolds
for Tissue Engineering**   55
S. Checa, C. Sandino, D.P. Byrne, D.J. Kelly, D. Lacroix
and P.J. Prendergast

**Modelling Bone Tissue Engineering. Towards an understanding
of the role of scaffold design parameters**   71
José A. Sanz-Herrera, Manuel Doblaré and José M. García-Aznar

**Geometric Modeling and Analysis of Bone Micro–Structures
as a Base for Scaffold Design**   91
Y. Holdstein, L. Podshivalov, A. Fischer

Electrospinning and Tissue Engineering   111
Geoffrey R. Mitchell and Fred Davis

Biofabrication Strategies for Tissue Engineering   137
Paulo Jorge Bártolo, Marco Domingos, Tatiana Patrício,
Stefania Cometa, Vladimir Mironov

Contents

Preface

Tissue Engineering is a multidisciplinary field involving scientists from different fields like medicine, chemistry, material science, engineering and biology. The development of mathematical methods is quite relevant to understand cell biology and human tissues as well to model, design and fabricate optimized and smart scaffolds.

The scientific interest of the computational mechanics community in this field, lead us to organize the First International ECCOMAS Thematic Conference on Tissue Engineering, held at the Centre for Rapid and Sustainable Product Development of the Polytechnic Institute of Leiria, Portugal, from July 9-11, 2009. It was a very successful event bringing together a considerable number of researchers from all over the world, representing several fields of study related to Tissue Engineering.

This conference had a number of distinguished keynote speakers that kindly agree to contribute to this book. The book consists of eight chapters describing the state of the art on computational modeling and fabrication in tissue engineering. We are deeply grateful to the all the contributing authors.

The Editors would also like to thank the European Community on Computational Methods in Applied Sciences (ECCOMAS), the Portuguese Association of Theoretical Applied and Computational Mechanics (APMTAC), the Portuguese Foundation for Science and Technology (FCT), the Institute of Mechanical Engineering (IDMEC/IST) and the Centre for Rapid and Sustainable Product Development of the Polytechnic Institute of Leiria (CDRsp), for supporting the Conference.

Paulo R. Fernandes
Paulo J. Bártolo

Cell mechanics: The role of simulation

Christopher R. Jacobs and Daniel J. Kelly

Abstract Computer simulation is one of the most powerful tools available to the applied mechanician to understand the complexities of mechanical behavior. It has revolutionized design of virtually all man-made structures from aircraft and buildings to cell phones and computers. It has also become a relatively important tool in biomechanics and simulation of tissues and implants has become routine. Indeed we appear to be on the verge of patient specific simulation becoming a critical tool in orthopaedic and cardiovascular surgery. However, its use as a tool of basic science is much less clear. In this chapter we explore the potential for mechanical simulation to contribute to improve fundamental understanding of biology. We consider the challenges of creating a model of a mechanobiological system with experimental validation. We propose that the area of cell mechanics is a particular area where simulation can make critically important contributions to understanding basic physiology and pathology and outline potential areas of future advancement.

1 Introduction

This manuscript is divided into three main sections. The topic of the first section was inspired by a Keynote lecture given at the 2007 European Society for Biomechanics Summer Workshop on Finite Element Analysis [1]. In this talk the author reviewed the role of simulation in making insights into biology and focused specifically on the capabilities and limitations of computer modeling to accomplish real hypothesis-testing-based investigation in biology. The contrast was made between hypothesis testing and descriptive research such as showing associations between mechanical and biological parameters or behavior. The conclusion is that one area where simulation might play an increasingly important role in identifying

Christopher R. Jacobs
Department of Biomedical Engineering, Columbia University, New York, NY, USA,
email: crj2111@columbia.edu

Daniel J. Kelly
Trinity Centre for Bioengineering, School of Engineering, Trinity College Dublin, Ireland.,
email: kellyd9@tcd.ie

P.R. Fernandes and P.J. Bártolo (eds.), *Advances on Modeling in Tissue Engineering*,
Computational Methods in Applied Sciences 20, DOI 10.1007/978-94-007-1254-6_1,
© Springer Science+Business Media B.V. 2011

biologic mechanistic behavior is cell mechanics. Indeed, it is for the most part difficult or impossible to experimentally quantify the mechanical state of a cell, suggesting that simulation can provide a valuable tool, particularly in the area of cell mechanobiology. Mechanobiology being the study of how biological systems sense and respond to mechanical signals in contrast to biomechanics, or the study of the mechanical behavior of biological systems [2]. We begin, in section 2, with a discussion of the potential and limitations of computer simulation of mechanical behavior. With this foundation, section 3 is a mini literature review of current work in cell mechanics, and section 4 is a brief discussion of exciting possibilities for the future.

2 What is the Role of Mechanical Simulation in cell Biology?

The finite element method has emerged as an incredibly powerful tool for numerical simulation of highly complex structures. It allows skilled engineers to accurately predict mechanical response incorporating aspects such as complex geometries, inhomogeneous materials, and complex material behavior such as anisotropy and non-linear behavior. It can also be used to simulate adaptive material behavior such as bone remodeling to specific loading conditions [3, 4]. In addition, modern simulation software has removed many of the time-consuming simulation tasks such as discretization and made simulation available to many more investigators than ever before. However, this ease of use also greatly increases the potential for misuse of the method and the prediction of inaccurate results. Furthermore, these erroneous predictions can appear to be quite authoritative due to sophisticated 3D graphical rendering software.

2.1 Verification, Validation, and Sensitivity

Anderson, Ellis, and Weiss [5] describe the process by which accuracy can be assured through considering three distinct aspects, verification, validation, and sensitivity analysis. The first, verification, involves ensuring that the software being employed is indeed computing a correct solution to the given problem, or correctly solving the equations. In other words, verification is ensuring that there are no errors in programming or bugs in the software package. Often in biomechanics, it is taken on faith that commercial software does in fact converge to a correct solution. Generally this is a safe assumption, however there are some infamous counter-examples [5]. More significant challenges result from issues of capturing singular behavior at non-smooth boundaries, ensuring adequate mesh refinement, non-linear material behavior, and properly dealing with contact behavior. In these cases it is simple to use a commercial code to obtain essentially meaningless results that certainly appear convincing [5] .

Assuming that the code and modeling strategy are all satisfactory, the next issue is the question of validation. This is the separate question of "to what extent the assumptions and parameters of the model reflect reality." Many times, in biomechanical analysis, it is not possible to consult a table of material properties as an analyst of traditional engineering structures might. Thus, the solution can only be expected to be a good as the assumptions made and GIGO (garbage in = garbage out) is a serious concern. The best case scenario is when the simulation can be compared to experimental validation data obtained by the same investigator or team [5]. A more common approach is to utilize data published from experiments done in another laboratory. The danger with this is that the experimental conditions are often not precisely similar between simulation and experiment, and in many cases the same material behavior parameter can differ wildly depending on experimental conditions. For example, cellular Young's modulus determined from atomic force microscopy is well known to be extremely sensitive to indenter tip geometry [6]. This implies that to have confidence in simulation results they should be validated with experimental results collected for the model output of interest. Validating a prediction of, for example, displacements do not ensure that force predictions are accurate. Indeed, the traditional finite element method is a displacement-based approach and displacements are accurate to a higher order than derived quantities such as stress or strain.

In reality, we are often faced with situations where one or more model parameters are not known with the precision we would like. This is often the situation with characterization of biological material. In this situation we must conduct a sensitivity or parametric analysis. The simulation can be conducted for a range of values for any unknown parameters, and the sensitivity of output determined. Confidence in the simulation may be gained if the model is insensitive to parameters that can only be determined to fall in some given range. If the simulation results do not change significantly as long as the input parameter falls within this range, there is no problem. In contrast, if the solution depends strongly on a parameter for which experimental data are not available, the specific predictions of the model may not be of much value. However, this situation can also be informative as it can focus experimentation on those parameters that are most critical. The bone remodeling simulation field is an interesting illustration of this principle. Often these simulations incorporate descriptions of biologic, mechanobiologic, and material behavior and interactions that can involved a large number of unknown parameters [7]. However, a parametric analysis suggests that reasonable bone morphologies are predicted regardless of the quantitative formulation adapted [8]. Although this is encouraging in terms of predicting bone behavior, it is also suggestive that it would be difficult to determine the quantitative relationship between bone physiology and mechanics with simulation using current modeling approaches.

2.2 The Paradox of Validation

In his presentation [1], the author introduced the concept of the paradox of validation. As described in 2.1, validation is a critical step creating a high-fidelity mechanical simulation. And validation should be conducted directly on the parameter of interest. In cell mechanics, for example, a simulation might be used to predict the membrane deformation or strain in a cell. This might be of interest to determine whether membrane deformation that occurs with particular physical signal such as fluid shear or substrate stretch is sufficient to activate a particular molecular signaling mechanism such as a stretch activated channel. This model would need to be validated with a measurement of membrane deformation through, for example, optical tracking of membrane-embedded protein. Alternatively, a simulation might be used to estimate the forces across a focal adhesion or even a single integrin [9] when a cell is loaded in a particular way. These predictions might be validated with a direct measurement using, for example, an atomic force microscope. The paradox is that the experimental data collected to validate the model may actually obviate the need for the model. In other words, by the time an experimental system is designed and built to measure cell-level mechanical behavior, it is likely just as easy to use experimental measurement directly than to rely on a model. Particularly since much of the time and money is making cell mechanics measurements is in designing and validating the rig.

A related issue in the application of simulation to cell mechanics is the perceived cost. Often the authors have encountered the assertion that simulation can generate more results than experimentation for a lower cost. However, it may be that this perception is a result of funding models in Europe. Although these models are changing, historically when investigators in certain institutions consider costs they do not necessarily include the cost of students, research staff, or their own time, which may not to be allocated to specific projects. Thus, the cost referred to is predominantly supply expenses, which are of course much lower for simulation. However, in the US model salaries are included in project costs and, indeed, typical represent 80% or more of total cost. This greatly increases the cost of simulation.

Another potential application of simulation in cell mechanics is to critically evaluate mechanical factors that drive biological process at a cell and molecular level and compare the system behavior with experimental data. This would be a great value when experimental strategies alone cannot address these relationships directly as well as to guide and leverage experimental investigations. For example, by comparing the patterns of differentiation in the regenerating tissue within loaded bone chambers with FE predictions of the biophysical environment, Lacroix and Prendergast hypothesized that shear strain and fluid flow regulated differentiation during bone regeneration [10]. It has been possible to continuously test this hypothesis by comparing the predictions of simulations based on the hypothesis with the outcomes of other regenerative events such as fracture healing, osteochondral defect repair and distraction osteogenesis [11-14]. In all cases the same underlying modeling parameters [15] were chosen to simulate tissue differentiation [10, 11, 13-15], so as to avoid the accusation of 'tweaking' model parameters to obtain better predictions. These modeling frameworks can also be used to critically compare different hypotheses for mechano-regulated progenitor cell differentiation, by comparing the predictive abilities of the different hypotheses [16]. Of course, some a priori insight is required to determine the initial set of model parameters. This was described by van der Meulen and Huiskes [2] as one of 'trial and error': "Computational mechanobiologists hypothesize a potential rule and determine if the outcome of this hypothesis produces realistic tissue structures and morphologies, hence 'trial-and-error'. If the results correspond well, they might be an explanation for the mechanism being modeled."

There are limitations to the above approaches. For example, models of mechanical regulation of tissue differentiation [10, 17] suggest that certain levels of shear strain and fluid flow acting on cells in a regenerating tissue will promote fibro-cartilaginous tissue formation. To date, these models have not been able to elucidate whether these cellular stimuli promote differentiation directly, or whether, for example, the mechanical environment is acting indirectly to promote chondrogenesis by inhibiting angiogenesis and thereby promoting the formation of a hypoxic environment known to promote chondrogenesis. Explicitly modeling the process of angiogenesis within these modeling frameworks may allow such questions to be more critically addressed [18].

Similar modeling approaches have been used in diverse areas of orthopedic science, from understanding the role of mechanics in development [19] to tissue organization during regeneration [20]. However, when comparing candidate cell-level mechanical signals it is important to remember the bone remodeling experience ([8] and 2.1 above) and critically evaluate the discriminatory power of comparing simulation results with histological or morphological data. Using a bone-ingrowth chamber Prendergast and colleagues have found that in order to evaluate falsifiable hypotheses concerning cell-level mechanical signals that regulate tissue differentiation a much larger set of animal-specific model with a high level of variability need to be considered than have been utilized in the past [21]. This is suggestive that perhaps important aspects of mechanoregulation in biology can be discerned with this approach when careful consideration is given to forming testable hypotheses, awareness of the relevant cell and molecular biology, and a critical level of biofidelity is achieved in the modeling.

3 State-of-the-art in Computational Cell Mechanics

Although there are important limitations and pitfalls to consider in simulating the mechanical behavior of cells, there have been important advances as well. In this section we review the recent literature regarding computational models of cell mechanics. The fundamental challenge that faces all of these efforts is how to approach the exceedingly complex internal structure of the cell and the diversity of behaviors which depend on cell type and morphology. We have grouped the review into homogenization approaches, microstructural models of the cytoskeleton, microstructural models of the cell membrane, and microstructural models of the entire cell with a focus on multi-scale modeling. By its nature this mini-review will only highlight important aspects of the field. More details treatments are already available in the literature [22].

3.1 Homogenization

For our purposes we define homogenization approaches to be those in which the microstructure of the cell is treated in a spatially averaged way with the goal of describing its apparent level behavior. Thus, these approaches include lumped-parameter and rheological models here. These approaches have been particularly valuable in the analysis and interpretation of local measurements of cellular behavior such as atomic force microscopy (AFM) and magnetic twisting cytometry.

Although simple in nature, treating red blood cells as elastic membranes surrounding a Maxwell viscous fluid can make surprisingly accurate predictions for the fluid properties of blood [23]. They have also been applied to understanding of leukocytes mechanics and adhesion [24] where they have been expanded to allow for large deformations [25]. This problem is made particularly challenging by the need to account for complex boundary conditions. For example, sliding must be included to simulation micropipette experiments needed for validation, but non-linear adaptive adhesion is required to simulate interaction with the vessel wall. The latter process is particularly complex and simulations have included effects of receptor/ligand density and affinity as well as cell motion and motility on the adhesive contact mechanics in both discrete and continuum approaches [26, 27]. Interestingly, it is known that the receptors that mediate extracellular adhesion, e.g. integrins, tend to function in clusters rather than discrete units [28]. The resulting simulation is highly nonlinear and involves two levels of coupling, that between cell deformation and adhesion as well as the fluid-structure interaction between cell deformation and fluid shear stress [25, 29-31].

Cell mechanics simulations have also been important in understanding structural tissues (i.e. tissues whose primary function is supporting or generating load). Simulations of smooth muscle cells have allowed investigators to make predictions of internal stress and strain [32]. The mechanical behavior of chondrocytes has been simulated with a multiphasic constitutive model. In this approach the viscoelastic material properties of the cell are deduced from the mechanical interaction of a hyperelastic cytoskeleton and the fluid cytoplasm flowing past it [33]. Chondrocyte models have been directly validated with AFM measurements and, interestingly, investigators were able to infer the mechanical behavior of the peri-cellular matrix as well as the cell itself [34].

An enclosed viscoelastic Maxwell fluid has also been utilized to simulate the mechanics of the nucleus [35]. However, although the cell's plasma membrane can be considered to be a simple elastic membrane with negligible bending stiffness, the nuclear membrane exhibits more complexity. The results of micropipette experiments on isolated nuclei have been accurately simulated with a three-layered nuclear membrane consisting of one elastic layer and two viscoelastic layers modeled rheologically as a standard linear solid.

Despite their success in blood cells and chondrocytes, homogenizing the cytoplasm/cytoskeleton fails to reflect the complex microstructure of the cytoskeletal polymer network that is important for many cell types. Critical structures such as the actin cortex or stress fibers are intracellular mechanical structures that are crucial to the cells mechanical performance. Of course, it would be impossible to fully review polymer network theory here, however, the diversity of cytoskeletal mechanics are of particular interest. For example, the polymers of the cytoskeleton imply that a single theory or approach is unlikely to yield a sufficient description. For example, the persistence length of microtubules is greater than a typical cell diameter. Thus, solid mechanics continuum models (e.g. simple columns or beams) may be appropriate. On the other hand, entropic effects can be responsible for the forces in actin microfilaments and intermediate filaments. However, when actin is highly cross-linked, its entropic behavior may be limited and a continuum description is again appropriate.

A homogenization strategy for cytoskeletal networks involves selecting a set of apparent level parameters to describe the network microstructure. For example volume fraction, orientation, and level of cross-linking might be appropriate. These parameters are then linked to apparent-level constitutive material behavior. Kwon et al. [36] presented a computational paradigm for achieving this homogenization and demonstrated it for actin networks of varying densities and anisotropies. The cytoskeleton has also been modeled as a fiber-reinforced composite and the predictions validated with whole-cell data from AFM and magnetic twisting cytometry [37]. Kamm and colleagues have recently presented a Brownian dynamics simulation of actin and actin cross-linkers to demonstrate regimes of entropy-dominated and energy-dominated behavior depending on polymer and crosslinker density as well as pre-stress [38].

3.2 The Cytoskeleton

The mechanics of the cytoskeleton are not only important in understanding cell mechanics and cytoskeleton-mediated processes such as motility and force generation, but also because the cytoskeleton is a site of mechanosensation [39] and the level of cytoskeletal pre-tension may regulate the cell's response to mechanical loading [40-42]. Microstructural models of the cytoskeleton often treat polymers as linear structures that can resist axial loads only. In the literature these systems are often referred to as tensegrity structures [43], although strictly speaking this is incorrect. Tensegrity systems incorporate only compression-only "struts" and tension-only "cables" and, in cells, microfilaments often supports compression and microtubules are frequently loaded in tension. Nonetheless, tensegrity models of the cytoskeleton have been shown to predict some of their dynamic properties [44]. A complete review of the diverse strategies for modeling cytoskeletal mechanics is presented in [45]. Additionally, microstructural models of the cytoskeleton are commonly a component of whole-cell microstructural models which will be discussed in the section 3.4. Intergrin mediated attachment of the cytoskeleton to the extracellular environment can also be critical to understanding cytoskeletal mechanics, not only as it relates to mechanical coupling, but also in terms of sensing mechanical loads and in the context of pathologies such as cancer [46].

3.3 The Cell Membrane

The cell membrane has perhaps received less attention as a mechanical structure than the cytoskeleton. Perhaps this is a result of a perception that the role of the cytoskeleton is primarily structural while the role of the membrane is primarily as a barrier. Indeed, this may be the case for the lipid bilayer. Due to the high mobility of the phospholipids that make up the bilayer, the shear modulus within the plane of the bilayer is negligible. As a consequence, its bending stiffness is also nearly zero. This accounts for the success of using the two dimensional membrane equation to describe its mechanics such as models described in section 3.1.

The low resistance of the bilayer to bending does, however, produce some interesting in-plane behavior. In many cell types and conditions, the cell's bilayer is slack state with in-plane stresses that are nearly zero. In this situation the membrane tends to take on an undulating configuration with a large number of folds or wrinkles. Under entropic Brownian influences, these folds can actually allow an effective or apparent tension to develop in the bilayer. Furthermore, this effect allows the bilayer to resist in-plane strain. Thus, the elasticity described in many membrane models originates entropically rather than from inter-molecular forces as in classical continuum mechanics. A separate mechanism of bilayer force generation occurs when it is stretched to the point that all of the folds have been removed. In this situation the hydrophobic interactions between the phospholipids resist further stretching and produce highly nonlinear behavior that approaches inextensibility. Indeed, the bilayer failure strain is on the order of 3%.

Although the bilayer had negligible bending stiffness, this is not necessarily true of the cell membrane. Often an underlying cytoskeletal network supports the bilayer. In red blood cells this is a specialize protein known as spectrin. In other cells actin provides bending resistance to the membrane. The resulting composite structure can have complex mechanical behavior including a significant transverse bending stiffness. A number of different continuum frameworks have been employed to describe cell membrane mechanics in simulations including classical shell theory, fluid mechanics, and statistical mechanics (reviewed in [47]). The mechanical interaction of the cytoskeleton and bilayer has been simulated in the context of actin polymers advancing the membrane during the extension of neurons. Also, since the since the bilayer is functionally a two dimensional fluid and only anchored to the member at discrete points, it is possible to pull cylindrical bilayer extensions known as tethers from a cell with functionalized micropipettes or AFM tips [48, 49]. Simulations of such processes can provide important validation of cell mechanics models.

3.4 Whole Cell Models and Multiscale Approaches

As microstructural models of cellular components such as the membrane and cytoskeleton have become more sophisticated, they have begun to be more frequently integrated to predicted whole cell mechanical behavior. For example, a whole-cell model of fibroblasts in suspension has been shown to predict optical tweezers data when details of the cortical actin are included [50]. Quantitative cell-cell forces have been determined in embryonic epithelial cells using a system of orthogonal dashpots to represent the viscoelastic cytoplasm/cytoskeleton [51]. Membrane blebbing (for formation of round protruding folds of membrane) has been simulated in whole-cell models that treat the membrane and cytoskeleton as elastic structures and an operator splitting approach to treat the membrane-cytoplasm fluid-structure interaction problem [52]. The bundles of cilia that act as mechanosensors in hair cells have been simulated in detailed finite element models that include cytoskeletal proteins, anchoring mechanics, bundle cross-linking, and tip links [53, 54]. Even the anisotropic deformation of yeast under extreme hydrostatic pressures (over 200MPa) have been simulated [55]. One challenge in whole-cell models is the complex microstucture of the cytoskeleton. The effect of the cytoskeleton has been incorporated into whole-cell models by simplifying its complex microstucture into a set of representative skeletal elements [56, 57]. This approach has also been applied to simpler spectrin network of the red blood cell [58].

Another strategy for microstructural simulation of entire structures is direct treatment of the microstructure. Highly efficient computational approaches have allowed for very large scale simulations of on the order of hundreds of millions of structural components. Such simulations have proven highly effective for treating microstucturally complex materials such as trabecular bone [59-61]. However, such approaches are insufficient to tackle the complexity of the cell at the filament or even fiber bundle level. As a result the only option for conducting whole-cell simulations with highly fidelic microstructures is to take a multi-scale approach. Although still in its infancy, some cell multi-scale simulations have been conducted. For example, Hartmann et al. recently simulated the deformation of a red blood cell deformed with optical tweezers with a multi-scale approach [62].

4 Future Directions

Of course, we can expect future simulations of cell mechanical behavior to rely more and more on multi-scale simulations. Interestingly, for cell mechanics this need not be limited to a simple matter of course-graining due to the size of the problem. In cell mechanics the challenge is inherently one of multi-physics as well. There is the obvious coupled fluid-structure problem represented by the cytoplasm and cytoskeleton/membrane respectively. However, other multi-physics issues are associated with adhesions and the cellular response and adaptation to mechanics as well as other phenomena not yet considered. This raises the potential that the ideal computational architecture for the different aspects of the problem might be quite different, as they are, for example, in fluid versus solid mechanics. Thus the application of heterogeneous computing approaches may play a critical role in the future of cell mechanics simulation. Indeed, cell mechanics could become one of the grand challenge problems in this new and exciting field.

Although multi-scale, multi-physics, and heterogeneous computing may be critical to future whole-cell mechanics simulations, there is also great potential for insight into fundamental cell behavior by modeling specific parts or components of cells. Models of the nucleus (described above) have led to insight into its architecture and mechanical behavior. The hair-cell cilia example described above is another example. Multi-cilia in the respiratory and female reproductive tracts are another. In this case the cilia are motile and molecular motors cause them to exhibit a wave-like beating motion that mobilize the extracellular fluid. Recently, computer simulations have been utilized to describe a fluid-structure coupling between multi-cilia and the bathing fluid that appears to be central to maintaining the coordination and progression of travelling wave [63].

Another area where very simple simulation may lead to important insights is in the mechanics of primary cilia. Primary cilia are related to multi-cilia and haircell cilia. However, they are nearly ubiquitous (only certain blood and kidney cells are known not to form primary cilia), every cell only has a single one, and until the last decade their function was largely a matter of speculation (Figure 1). However, recently, a rapidly increasing body of evidence suggests that primary cilia act as extracellular sensors, sampling the biomechanical and biochemical environment some distance away from the cell [64, 65]. In our laboratory we are focused on understanding the ability of primary cilia to sense fluid flow. Specifically, this is likely related to the deformation that the cilium experiences. We have recently constructed a simple-but-effective large-rotation beam-bending model of the cilium that allows for rotation of the intracellular anchorage within the cilium (Figure 2). Interestingly, when rotation was not allowed the predicted flexural rigidity of the model was half the value obtain when rotation was allowed.

Multi-scale modeling may also overcome many of the limitations that have impeded the ability of simulation to truly accomplish hypothesis-testing investigation. We previously referred to the bone remodeling simulation field, and the observation that reasonable bone morphologies could be predicted for many quantitative formulations (7). By combining different hierarchal levels of experimental data, it may be possible to use multi-scale modeling to ask not if different magnitudes of stress, strain or strain energy density at the tissue level regulate bone remodeling, but rather if alterations in tissue level loading can lead to deformation in specific cellular structures or even to conformational changes in particular proteins. In this way it may be possible for modelers to explain how the observations of experimentalists at the cell or molecular level can result in changes observed at a tissue level during development, aging and disease.

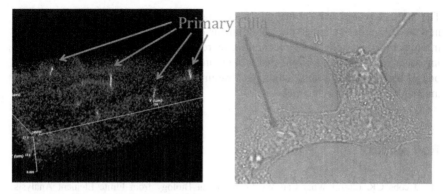

Fig. 1: Primary cilia in cultured kidney cells imaged by confocal microscopy (left) and epifluorescence (right).

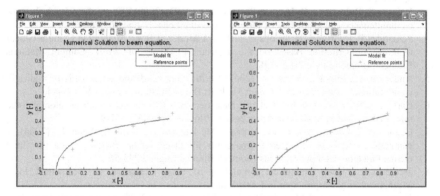

Fig. 2: We validated our simple model of primary cilium bending with taken from the literature for kidney cells in culture [66]. Interestingly, when basal twisting is not allowed (left) the predicted flexural rigidity is half as large as when it is allowed (right).

5 Conclusions

The mechanical behavior of cells is one of the most significant challenges facing biomechanics today. It is at the heart of a number of diseases responsible for great human suffering. It is also central to understanding cellular mechantransduction (particularly in non-excitable cells) and the mechanobiology of structural tissues such as bone and cartilage. Yet despite this critical importance, it receives dramatically less attention than traditional biologic and biochemical research. Perhaps this is due to the scarcity of individuals trained with deep understanding in both engineering mechanics simulation and modern cell and molecular biology.

The authors would like to strongly encourage work in this area and particularly new investigators consider the potential for critical new insight by pursuing this interdisciplinary field. Cell mechanics represents not only one of the highest-impact areas the young biomechanical engineer can choose to work in, but also one of the most demanding challenges requiring facility in the state-of-the-art in mechanical simulation technology.

References

1. Jacobs CR, (2007) What Can We Learn About Biology from Finite Element Analysis. European Society for Biomechanics Summer Workshop. Trinity Collage Dublin.
2. Van der Meulen MC, Huiskes R (2002) Why mechanobiology? A survey article. J Biomech 35:401-14.
3. Huiskes R, Ruimerman R, Van Lenthe GH, Janssen JD, (2000) Effects of mechanical forces on maintenance and adaptation of form in trabecular bone. Nature 405:704-6.
4. Fernandes P, Rodrigues H, Jacobs C, (1999) A Model of Bone Adaptation Using a Global Optimisation Criterion Based on the Trajectorial Theory of Wolff. Comput Methods Biomech Biomed Engin 2:125-138.
5. Anderson AE, Ellis BJ, Weiss J A, (2007) Verification, validation and sensitivity studies in computational biomechanics. Comput Methods Biomech Biomed Engin 10:171-84.
6. Carl P, Schillers H, (2008) Elasticity measurement of living cells with an atomic force microscope: data acquisition and processing. Pflugers Arch 457: 551-9.
7. Cegonino J, Garcia Aznar JM, Doblare M, Palanca D, Seral B, Seral F, (2004) A comparative analysis of different treatments for distal femur fractures using the finite element method. Comput Methods Biomech Biomed Engin 7:245-56.
8. Jacobs CR, (1994) Numerical simulation of bone adaptation to mechanical load. Mechanical Engineering. Stanford, Stanford.
9. Wang Y, McNamara LM, Schaffler MB, Weinbaum S, (2008) Strain amplification and integrin based signaling in osteocytes. J Musculoskelet Neuronal Interact 8:332-4.
10. Lacroix D, Prendergast PJ (2002) A mechano-regulation model for tissue differentiation during fracture healing: analysis of gap size and loading. J Biomech 35:1163-71.
11. Kelly DJ, Prendergast PJ, (2005) Mechano-regulation of stem cell differentiation and tissue regeneration in osteochondral defects. J Biomech 38:1413-22.
12. Khayyeri H, Checa S, Tagil M, Prendergast PJ (2009) Corroboration of mechanobiological simulations of tissue differentiation in an *in vivo* bone chamber using a lattice-modeling approach. J Orthop Res 27:1659-1666.
13. Andreykiv A, Van Keulen F, Prendergast PJ, (2008) Simulation of fracture healing incorporating mechanoregulation of tissue differentiation and dispersal/proliferation of cells. Biomech Model Mechanobiol 7:443-61.
14. Boccaccio A, Prendergast PJ, Pappalettere C, Kelly DJ, (2008) Tissue differentiation and bone regeneration in an osteotomized mandible: a computational analysis of the latency period. Med Biol Eng Comput 46:283-98.
15. Huiskes R, Van Driel WD, Prendergast PJ, Soballe K, (1997) A biomechanical regulatory model for periprosthetic fibrous-tissue differentiation. J Mater Sci Mater Med, 8:785-8.
16. Isaksson H, Wilson W, Van Donkelaar CC, Huiskes R, Ito K, (2006) Comparison of biophysical stimuli for mechano-regulation of tissue differentiation during fracture healing. J Biomech 39:1507-16.
17. Hayward LN, Morgan EF, (2009) Assessment of a mechano-regulation theory of skeletal tissue differentiation in an *in vivo* model of mechanically induced cartilage formation. Biomech Model Mechanobiol 8:447-455.

18. Checa S, Prendergast PJ, (2009) A mechanobiological model for tissue differentiation that includes angiogenesis: a lattice-based modeling approach. Ann Biomed Eng 37:129-45.
19. Nowlan NC, Murphy P, Prendergast PJ, (2007) Mechanobiology of embryonic limb development. Ann N Y Acad Sci 1101:389-411.
20. Nagel T, Kelly DJ, (2010) Mechano-regulation of mesenchymal stem cell differentiation and collagen organisation during skeletal tissue repair. Biomech Model Mechanobiol, 9, 359-72.
21. Khayyeri H, (2010) in Jacobs CR (Ed.). Valencia, Spain.
22. Vaziri A, Gopinath A, (2008) Cell and biomolecular mechanics in silico. Nat Mater 7:15-23.
23. Skalak R, Keller SR, Secomb TW, (1981) Mechanics of blood flow. J Biomech Eng 103:102-15.
24. Skalak R, Dong C, Zhu C, (1990) Passive deformations and active motions of leukocytes. J Biomech Eng 112:295-302.
25. Dong C, Skalak R, (1992) Leukocyte deformability: finite element modeling of large viscoelastic deformation. J Theor Biol 158:173-93.
26. DiMilla PA, Barbee K, Lauffenburger DA, (1991) Mathematical model for the effects of adhesion and mechanics on cell migration speed. Biophys J 60:15-37.
27. N'Dri NA, Shyy W, Tran-Son-Tay R (2003) Computational modeling of cell adhesion and movement using a continuum-kinetics approach. Biophys J 85:2273-86.
28. Ward MD, Dembo M, Hammer DA, (1994) Kinetics of cell detachment: peeling of discrete receptor clusters. Biophys J 67:2522-34.
29. Dong C, Lei XX, (2000) Biomechanics of cell rolling: shear flow, cell-surface adhesion, and cell deformability. J Biomech 33:35-43.
30. Kan HC, Udaykumar HS, Shyy W, Tran-Son-Tay R, (1999) Numerical analysis of the deformation of an adherent drop under shear flow. J Biomech Eng 121:160-9.
31. Pozrikidis C, (2003) Numerical simulation of the flow-induced deformation of red blood cells. Ann Biomed Eng 31:1194-205.
32. Bursa J, Lebis R, Janicek P, (2006) FE models of stress-strain states in vascular smooth muscle cell. Technol Health Care 14:311-20.
33. Wu JZ, Herzog W, (2000) Finite element simulation of location- and time-dependent mechanical behavior of chondrocytes in unconfined compression tests. Ann Biomed Eng 28:318-30.
34. Ng L, Hung HH, Sprunt A, Chubinskaya S, Ortiz C, Grodzinsky A (2007) Nanomechanical properties of individual chondrocytes and their developing growth factor-stimulated pericellular matrix. J Biomech 40:1011-23.
35. Vaziri A, Mofrad MR, (2007) Mechanics and deformation of the nucleus in micropipette aspiration experiment. J Biomech 40:2053-62.
36. Kwon RY, Lew AJ, Jacobs CR, (2008) A microstructurally informed model for the mechanical response of three-dimensional actin networks. Comput Methods Biomech Biomed Engin 11:407-18.
37. Unnikrishnan GU, Unnikrishnan VU, Reddy JN, (2007) Constitutive material modeling of cell: a micromechanics approach. J Biomech Eng 129:315-23.
38. Kim T, Hwang W, Lee H, Kamm RD, (2009) Computational analysis of viscoelastic properties of crosslinked actin networks. PLoS Comput Biol 5: e1000439.
39. Colombelli J, Besser A, Kress H, Reynaud EG, Girard P, Caussinus E, Haselmann U, Small JV, Schwarz US, Stelzer, EH, (2009) Mechanosensing in actin stress fibers revealed by a close correlation between force and protein localization. J Cell Sci 122:1665-79.
40. Jaasma MJ, Jackson WM, Tang RY, Keaveny TM, (2007) Adaptation of cellular mechanical behavior to mechanical loading for osteoblastic cells. J Biomech 40:1938-45.
41. You L, Temiyasathit S, Coyer SR, Garcia AJ, (2008) Bone cells grown on micropatterned surfaces are more sensitive to fluid shear stress. Cell and Molecular Bioengineering 1:182-188.

42. Arnsdorf EJ, Tummala P, Kwon RY, Jacobs CR, (2009) Mechanically induced osteogenic differentiation–the role of RhoA, ROCKII and cytoskeletal dynamics. J Cell Sci 122:546-53.
43. Ingber DE, (1998) The architecture of life. Sci Am, 278:48-57.
44. Sultan C, Stamenovic D, Ingber DE, (2004) A computational tensegrity model predicts dynamic rheological behaviors in living cells. Ann Biomed Eng, 32:520-30.
45. Volokh KY, (2003) Cytoskeletal architecture and mechanical behavior of living cells. Biorheology, 40:213-20.
46. Baker EL, Zaman MH, (2009) The biomechanical integrin. J Biomech 43:38-44.
47. Brown FL, (2008) Elastic modeling of biomembranes and lipid bilayers. Annu Rev Phys Chem, 59:685-712.
48. Allen KB, Sasoglu FM, LAYTON BE, (2009) Cytoskeleton-membrane interactions in neuronal growth cones: a finite analysis study. J Biomech Eng, 131:021006.
49. Schumacher KR, Popel AS, Anvari B, Brownell WE, Spector AA, (2008) Modeling the mechanics of tethers pulled from the cochlear outer hair cell membrane. J Biomech Eng 130:031007.
50. Ananthakrishnan R, Guck J, Wottawah F, Schinkinger S, Lincoln B, Romeyke M, Moon T, Kas J, (2006) Quantifying the contribution of actin networks to the elastic strength of fibroblasts. J Theor Biol 242:502-516.
51. Brodland GW, Viens D, Veldhuis JH, (2007) A new cell-based FE model for the mechanics of embryonic epithelia. Comput Methods Biomech Biomed Engin 10:121-8.
52. Young J, Mitran S, (2010) A numerical model of cellular blebbing: A volume-conserving, fluid-structure interaction model of the entire cell. J Biomech. 42:210:20.
53. Duncan RK, Grant JW, (1997) A finite-element model of inner ear hair bundle micromechanics. Hear Res 104:15-26.
54. Cotton JR, Grant JW, (2000) A finite element method for mechanical response of hair cell ciliary bundles. J Biomech Eng 122:44-50.
55. Hartmann C, Delgado A, (2004) Numerical simulation of the mechanics of a yeast cell under high hydrostatic pressure. J Biomech 37:977-87.
56. McGarry JG, Klein-Nulend J, Mullender MG, Prendergast PJ, (2005) A comparison of strain and fluid shear stress in stimulating bone cell responses–a computational and experimental study. FASEB J 19:482-4.
57. Charras GT, Horton MA, (2002) Determination of cellular strains by combined atomic force microscopy and finite element modeling. Biophys J 83:858-79.
58. Boey SK, Boal DH, Discher DE, (1998) Simulations of the erythrocyte cytoskeleton at large deformation. I. Microscopic models. Biophys J 75:1573-83.
59. Bevill G, Keaveny TM, (2009) Trabecular bone strength predictions using finite element analysis of micro-scale images at limited spatial resolution. Bone 44:579-84.
60. Pahr DH, Zysset PK, (2009) A comparison of enhanced continuum FE with micro FE models of human vertebral bodies. J Biomech 42:455-62.
61. Van Rietbergen B, Muller R, Ulrich D, Ruegsegger P, Huiskes R, (1999) Tissue stresses and strain in trabeculae of a canine proximal femur can be quantified from computer reconstructions. J Biomech 32:443-51.
62. Hartmann D, (2010) A multiscale model for red blood cell mechanics. Biomech Model Mechanobiol 9:1-17.
63. Dillon RH, Fauci LJ, (2000) An integrative model of internal axoneme mechanics and external fluid dynamics in ciliary beating. J Theor Biol 207:415-30.
64. Malone AM, Anderson CT, Tummala P, Kwon RY, Johnston TR, Stearns T, Jacobs CR, (2007) Primary cilia mediate mechanosensing in bone cells by a calcium-independent mechanism. Proc Natl Acad Sci U S A 104:13325-30.
65. Anderson CT, Castillo AB, Brugmann SA, Helms JA, Jacobs CR, Stearns T, (2008) Primary cilia: cellular sensors for the skeleton. Anat Rec (Hoboken) 291:1074-8.
66. Schwartz EA, Leonard ML, Bizios R, Bowser SS, (1997) Analysis and modeling of the primary cilium bending response to fluid shear. Am J Physiol 272: F132-8.

Multiscale Modelling of Bone Tissue – Remodelling and Application to Scaffold Design

Helder C. Rodrigues, Pedro G. Coelho, Paulo R. Fernandes

Abstract In tissue engineering design it is in general hypothesised that the scaffold should provide a mechanical environment similar to the pre-degenerative one for initial function and have sufficient pore interconnectivity for cell migration and cell/gene delivery. In the case of bone tissue, the design of such scaffolds can be greatly improved by the knowledge of the bone adaptation process thus facilitating the identification of scaffold microstructures compatible with the properties of real bone. In this chapter we present a multiscale model for bone adaptation that potentially can respond to some of the scaffold design requirements. The proposed model focuses on the two top scale levels of bone architecture. The macroscale (whole bone) characterized by the bone apparent density distribution and the microscale where the trabecular structure of bone in terms of its mechanical properties is characterized. At global scale bone is assumed as a continuum material characterized by equivalent mechanical properties. At local scale the bone trabecular anisotropy is approached by a locally periodic porous material. The relevance of incorporating a micro design scale relies on the possibility of controlling morphometric parameters that not only characterize trabecular structure, and thus can help the design of bone substitutes, but also allow a fine balance between bone tissue biological and mechanical functions.

Helder C. Rodrigues
IDMEC – Instituto Superior Técnico, TU Lisbon, Portugal
email: hcr@ist.utl.pt

Pedro G. Coelho
Mechanical and Industrial Engineering Department, FCT/UNL, Portugal
email: pgc@fct.unl.pt

Paulo R. Fernandes
IDMEC – Instituto Superior Técnico, TU Lisbon, Portugal
email: prfernan@dem.ist.utl.pt

P.R. Fernandes and P.J. Bártolo (eds.), *Advances on Modeling in Tissue Engineering,*
Computational Methods in Applied Sciences 20, DOI 10.1007/978-94-007-1254-6_2,
© Springer Science+Business Media B.V. 2011

1 Introduction

Bone is a natural porous material that presents a well defined hierarchical structure suggesting a biological functional adaptation to the mechanical environment under strict biological requirements. For instance, if we observe the structure of a long bone, bone tissue is dense on the external boundaries (cortical type) and is porous within the epiphyses (cancellous type), as a result of the balance between the functional demands, both structural (external loading) and biological (cell nutrition, vascularisation, mineral and marrow fat reservoir, etc). This functional gradation inspires the use of functional graded materials for biomechanical devices such as bone implants and bone substitutes or scaffolds [1]. In this context a proper mathematical characterization of the bone functional adaptation process is a critical step in the design of new bone implants and bone scaffolds for tissue engineering [2].

Bone adaptation depends on several factors from which the local mechanical environment plays a crucial role. Since Wolff's observations, abridged in what is generally know as Wolff's Law [3], numerous mathematical and computational models for bone functional adaptation have been developed [4-16]. An important issue of these models is how to model the human bone as a structural material and how its mechanical properties are obtained. Some of the earlier works assumed bone as an isotropic material [4-8]. Later ones, such as the ones presented by Jacobs et al. [9], Fernandes et al. [13] and Doblaré and Garcia et al. [15], consider the anisotropy of bone coupling apparent density and orientation into a single model. Within this later group we can distinguish the class of multiscale models that have a special interest for the work presented in this chapter (see *e.g.* Rodrigues et al. [14] and Coelho et al. [16]). These models go beyond the description of the anisotropic material properties for each bone anatomical site. Actually, they can define the anisotropic characteristics of the microstructure for each bone anatomical site and associate with it a specific (even though not unique) microstructure.

Based on this multiscale concept for bone adaptation, we describe here a model to support the design of scaffolds for bone tissue engineering. The model considers the two top of levels of bone structure [16]: the macroscopic level where the bone apparent density is characterized and a microscopic level where the trabecular structure is defined by its mechanical properties. The mathematical law of bone remodelling states that bone adapts to functional demands in order to satisfy a multi-criteria balancing structural stiffness and metabolic cost of bone formation. The multiscale model presented, when combined with suitable fabrication techniques and biomaterials, provides a valuable tool to obtain three-dimensional porous substrates exhibiting controlled internal micro-architectures that favour living cell culture and migration. Thus, this chapter addresses also the production of bone scaffolds based on designs obtained with the proposed hierarchical material optimization model.

2 Bone remodelling multiscale Problem

Bone presents a hierarchical structure, from nano-scale to macro-scale, where it is possible to identify different levels of structural organization [17]. The two top levels are the whole bone and the trabecular architecture of spongy bone. In the multi-scale approach presented here, it is assumed that bone is a porous (cellular) locally periodic material as shown in figure 1. The domain Ω, corresponding to the whole bone, is defined for a macro-scale level while the design domain Y is defined for a micro-scale level, understood here as being the trabecular scale. The model computes not only the relative bone volume fraction ρ in each point $\mathbf{x}$ of Ω (global scale or macroscale) but also the volume fraction μ in each point $\mathbf{y}$ of Y (local scale or microscale). Therefore, two material distribution problems can be identified, a global and a local one. The local problem in Y defines the topology of a single cell (trabecular bone base cell) which is assumed to be periodically repeated inside a small neighborhood of a point $\mathbf{x} \in \Omega$. In this material model, the relative density ρ in the macroscale is coupled with the density μ of microscale by the following equation,

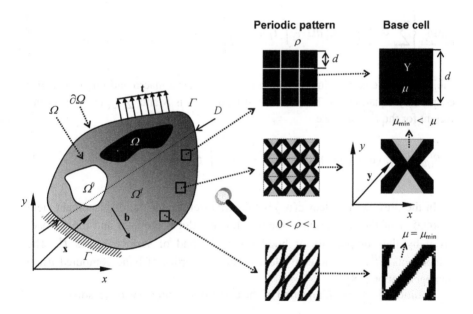

Fig. 1: Multi-scale bone remodelling model. The whole bone domain Ω is subject to external loads t on Γ_t, body forces b on Ω and it is fixed on Γ_u.

$$\rho(\mathbf{x}) = \frac{1}{|Y|} \int_Y \mu(\mathbf{x}, \mathbf{y}) dY, \quad \forall \mathbf{x} \in \Omega; \ 0 < \mu(\mathbf{x}, \mathbf{y}) \le 1 \tag{1}$$

for every point $\mathbf{x}$ of the global problem, where $|Y|$ is the volume of the base cell domain.

As a first step, let us assume that bone adapts in response to mechanical demands with the bone mass regulated by the metabolic cost of bone apposition. Following this assumption the law of bone remodelling can be defined as an optimization problem linking two conflicting objectives: maximization of material stiffness and minimization of the cost of bone remodelling. The assumption here is that bone presents maximum mechanical efficiency with minimum mass. Furthermore, since the time scale of applied force variations is much smaller than the bone remodelling time scale we will characterize the applied forces as a weighted set of multiple equivalent static loads (multiload) characterizing the loading environment applied to the bone, instead of a unique varying load function of time "t".

Thus, assuming the multiload formulation, the stiffness maximization leads to the following minimization problem,

$$\min_{\rho, 0 \le \mu \le 1} \left\{ \sum_{r=1}^{N} \alpha^r \left(\int_\Omega b_i^r u_i^r d\Omega + \int_\Gamma t_i^r u_i^r d\Gamma \right) \right\} \tag{2}$$

subjected to the equilibrium equation (3) and with the macro and micro densities coupled by equation (1). The number of load cases is N and $\sum_{r=1}^{N} \alpha^r = 1$ is the weight for the rth load case.

$$\int_\Omega E_{ijkl}^H(\mu)\varepsilon_{ij}(\mathbf{u}^r)\varepsilon_{kl}(\mathbf{v}^r)d\Omega - \int_\Omega b_i^r v_i^r d\Omega - \int_\Gamma t_i^r v_i^r d\Gamma = 0, \forall \mathbf{v}^r \text{ adm.} \tag{3}$$

In the previous problem $\varepsilon(\mathbf{u}^r)$ is the strain tensor for the displacement field $\mathbf{u}^r$ solution of the equilibrium equation for the global problem and the rth load, given by the components of body forces b_i^r and of surface forces t_i^r. The components E_{ijkl}^H are the equivalent elastic properties of bone computed using homogenization [18].

Additionally, the minimization of the cost k of creating new bone solves,

$$\min_{\rho, 0 \le \mu \le 1} \left\{ \int_\Omega k\rho \, d\Omega \right\} \tag{4}$$

With $\rho(x)$ given by equation (1).

From stationary condition of the objective functional the solution of the above minimization problems (2-4) is characterized by the necessary condition,

$$\frac{\partial E_{ijkl}^{H}(\mu)}{\partial \mu(\mathbf{x}, \mathbf{y})} \sum_{r=1}^{N} \left[\alpha^{r} \; \varepsilon_{ij}(\mathbf{u}^{r}) \varepsilon_{kl}(\mathbf{v}^{r}) \right] - k = 0 \tag{5}$$

Equation (5) is the "law of bone remodelling" in the sense that whenever it is satisfied for intermediate values of μ the remodelling equilibrium is achieved. Thus the "micro" density field μ changes until this equation is satisfied.

An alternative formulation for the previous optimization problem is (see Rodrigues et al. [14,19] and Bendsoe and Sigmund [20]),

$$\min_{\substack{\rho(x) \\ \rho_{min} \leq \rho(x) \leq \rho_{max}}} \left[\int_{\Omega} \Phi(\rho, \mathbf{u}^{1}, \dots, \mathbf{u}^{P}) \, d\Omega + k \int_{\Omega} \rho(x) \, d\Omega \right] \tag{6}$$

$$\Phi(\rho, \mathbf{u}^{1}, \dots, \mathbf{u}^{P}) = \max_{\substack{\mu(y) \\ \mu_{min} \leq \mu(y) \leq 1 \\ \int_{Y} \mu(y) dY = \rho(x)}} \sum_{r=1}^{P} \alpha^{r} \left[\frac{1}{2} E_{ijkl}^{H}(\mu) \varepsilon_{ij}(\mathbf{u}^{r}) \varepsilon_{kl}(\mathbf{u}^{r}) \right] \tag{7}$$

where $\varepsilon(\mathbf{u}^{r})$ is the strain tensor for the displacement field $\mathbf{u}^{r}$ solution of the equilibrium equation for the global problem and the rth load, and E_{ijkl}^{H} is the homogenized properties for the trabecular bone at a point $x \in \Omega$ [18]. The parameter α^{r} is the weight for the rth load case and $\sum_{r=1}^{P} \alpha^{r} = 1$. Parameter k is the metabolic cost of bone apposition defined above.

This formulation clearly shows the multi-scale nature of the problem being solved, identifying two material distribution problems; a global one governed by ρ (eq. 6) and a local one governed by μ (eq.7).

3 Computational Model

Usually the problem above is solved simply by the fixed point type iterative scheme to compute de micro density μ, while ρ is *a posteriori* given by (1).

Note that two finite element (FE) meshes are required, one to approximate the macroscale problem and another the microscale problem (see e.g. references [16] and [24] for further details).

Thus assuming that ρ^E and μ_e^E, (macro and micro densities respectively) constant within each Ω^E and Y^e finite element, where the indexes e ($e =1,...,m$) and E ($E =1,...,M$) refer to the finite the local and global elements respectively, the fixed point iterative algorithm is,

$$\mu_{i+1}^e = \begin{cases} \max[(1 - \zeta)\mu_i^e, \mu_{min}] & if \ \ \gamma^e\mu_i^e \leq \max[(1 - \zeta)\mu_i^e, \mu_{min}] \\ \gamma^e\mu_i^e \ if \ \max[(1 - \zeta)\mu_i^e, \mu_{min}] \leq \ \gamma^e\mu_i^e \leq \min[(1 + \zeta)\mu_i^e, 1] \\ \min[(1 + \zeta)\mu_i^e, 1] & if \ \ \gamma^e\mu_i^e \leq \min[(1 + \zeta)\mu_i^e, 1] \end{cases} \quad (8)$$

$$\gamma^e = \frac{\sum_{r=1}^N \left[\alpha^r \frac{\partial E^H(\mu)}{\partial \mu(y)} \langle \varepsilon(u^r) \rangle_E \langle \varepsilon(u^r) \rangle_E \right]}{k} \quad (9)$$

In the above algorithm the parameter k, in the denominator, is the cost of bone apposition that depends on biological factors such as: disease, age, gender, and nutrition. The factor ζ is a numerical stabilization parameter for the density update.

The remodelling stimulus γ (8) is based on the actual displacement field and respective strain field for each load case "r". For this the equilibrium equation, in virtual work form (3), is solved using the Finite Element Method (FEM). The density ρ is assumed constant within each macro finite element (denoted by "E" in eqs. 8-9) and the local density μ is also assumed constant within each micro finite element (denoted by "e" in eqs. 8-9).

However in the context of bone adaptation, simulations based on (1), (8) and (9) are not the best way to introduce biological driven information such as bone surface density or bone permeability [16]. In fact, the proper way to introduce this information into the model is to define additional constraints on the microstructure design to approach biological requirements of actual bone. Moreover, the above presented iterative scheme leads to cortical (compact) bone identified with full material solutions (isotropy case), i.e., all μ's = 1 or ρ = 1. Actually compact bone is not completely solid but instead a porous medium with low porosity, around 5%–10% [21].

In order to obtain a better approach in this regard it is possible to set an upper bound ($\rho_{max} < 1$) on the apparent bone density in the macro problem to model the porosity of cortical bone (see eq. 6). In order to handle this additional constraint and solve problems (6) and (7) there are some algorithmic strategies much more efficient than the iterative scheme presented. Such strategies combine two optimizers, taking each one the respective density based design variable ρ or μ (see Coelho et al. [16]).

To solve the multiscale problem defined by (6) and (7), the Method of Moving Asymptotes – MMA [22] is combined with CONvex LINearization scheme (CONLIN), as a particular case of MMA [23] to solve (6) on ρ and (7) on μ, respectively. Figure 2 describe the working of the algorithm by means of a flowchart. Here it is highlighted the parallel version of the algorithm quite appropriate due to the independence property of local problems (see Coelho et al. [24] for details).

It should be noted that there are as many local problems as the number of global finite elements since associated to each global element Ω^E there is a local pattern generated by the repetition of a single base cell which is designed in Y. The parallelization technique used here splits the inner loop among several processors such that each processor solves a package of local problems corresponding to the microstructure characterization inside some macro finite elements. The main data interchanged between local and global problems is shown in the figure 2 through arrows with dashed lines. Summarizing, the global problem supplies each local problem with respective strains and macro density. In turn, local problems give back to global one the homogenized elastic properties and the Lagrange multiplier associated to local volume fraction constraint (1) which plays a role in the sensitivity of the global problem [24].

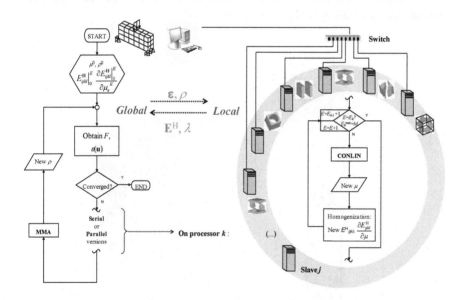

Fig. 2: Flow chart for the computational multiscale procedure. The global problem sends to the local one the strain field and the density for each element. The local problem gives back the bone equivalent properties (homogenized) and the Lagrange multiplier necessary to compute ρ.

4 Multiscale modelling of the femur bone

The multiscale material model was applied to the femur bone. A finite element model of a proximal human femur was developed. The bone geometry is based on the standardized femur [25]. The femur bone domain Ω and cell domain Y are discretized by 2112 and 8000 (20×20×20) hexahedral isoparametric finite elements of 8 nodes, respectively (see figure 3). The value for k was assumed equal to 1050 with correspond to a global bone mass of 50%.

A multiload approach was considered based on the load data published by Bergmann et al. [26, 27]. Ten load cases were considered (P = 10) [28]. A first set of five load cases corresponds to normal walking activity and a second one corresponds to stair climbing. All load cases have equivalent weight ($\alpha = 0.1$) [28].

It is assumed an isotropic base material for bone tissue defined by 20 GPa of Young's Modulus and 0.3 of Poisson's ratio. It should be noticed that the resulting trabecular bone appears as a combination of dense mineralized struts (base material) and pores, while the predicted structure for compact bone is a mineralized matrix (base material) crossed by canals ("haversian canals"). Once a micro density distribution is known, the bone equivalent properties, both for trabecular and compact bone, are obtained by homogenization.

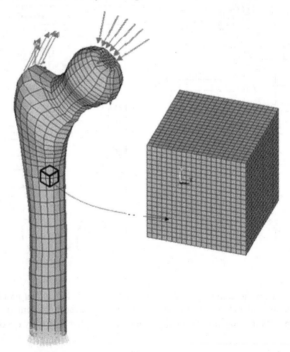

Fig. 3: Finite Element meshes for the macroscale problem (whole bone) and for the microscale problem (microstructure). Both meshes are made of 8 node hexahedral elements. The macoscale mesh has 2112 elements and the microscale mesh has 8000 elements. Arrows are a schematic representation of the applied loads.

The results for apparent density distribution are shown in figure 4, where black correspond to compact bone and grey to spongy bone, void is empty space. This distribution reproduces some characteristics of the real femur such as the external compact bone layer, the medullar cavity and the less dense spongy bone in the femoral head, with the Ward's triangle clearly identified.

The microstructures representative of trabecular bone, for different anatomic regions, are shown in Figures 5, 6 and 7. In figure 5, the microstructures corresponding to cortical bone are represented. The cortical bone is not modelled as a full material solution (isotropic) but instead as a porous medium matching real bone porosity (5%–10%, see Martin et al. [21]). In this example a $\rho_{max} = 0.925$ is used to assure a minimum of porosity of 7,5% in cortical bone. The porosity limit of $\rho_{max} = 0.925$ results on longitudinal cavities resembling Haversian canals, which are actually the main contribution for cortical bone porosity and anisotropy (transversal isotropy).

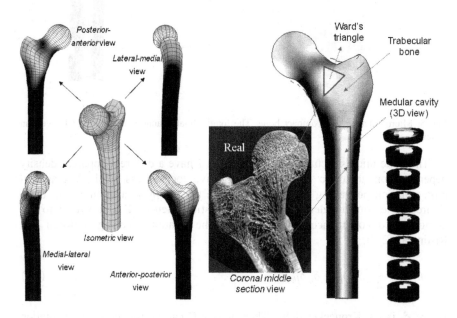

Fig. 4: Bone apparent density distribution. Compact bone is black while grey represents trabecular bone. The medullar cavity and the ward triangle are clearly identified.

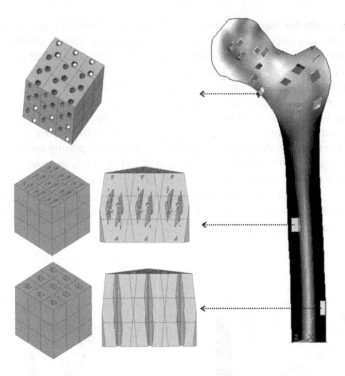

Fig. 5: Microstructures for compact bone. The longitudinal channels resemble the Harversian system.

The bone microstructures in figures 6 and 7 have a different apparent density depending on anatomic region and respective local stress field. A graphical representation portraying the trabecular bone anisotropy is also shown in these figures. The fourth-order homogenized elasticity tensor $\mathbf{E}$ is inverted to get homogenized compliance tensor $\mathbf{C}$ using the relation (see e.g. Ottosen and Ristinmaa [29]),

$$E_{ijkl}C_{ijkl} = \frac{1}{2}\left(\delta_{ik}\delta_{jl} + \delta_{il}\delta_{jk}\right) \tag{10}$$

where δ_{ij} is the Kronecker delta. Then, tensor $\mathbf{C}$ can be rotated using a rotation tensor $\mathbf{R}$ in terms of Euler angles,

$$C'_{ijkl} = R_{im}R_{jn}R_{kp}R_{lq}C_{mnpq} \tag{11}$$

At each orientation the component $1/C'_{1111}$ has the meaning of a directional stiffness and is plotted as function of orientation. For isotropic microstructures, it is simply the Young's Modulus and the resulting plot is a sphere.

In regions where a triaxial state of stress is dominant the microstructures are equiaxed (closest to the isotropic behavior) as shown in figure 6. There are regions where microstructures are strongly oriented (Figure 7). For uniaxial stress field and biaxial the resulting microstructures become like honeycomb or parallel plates structures, respectively. These results agree with idealized bone microstructures as presented by several authors (see e.g. Gibson and Ashby [30]).

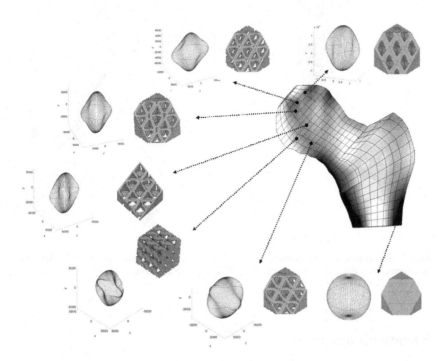

Fig. 6: Microstructures for trabecular bone. The bone regions presented in this figure show microstructures with material symmetry closer to the isotropic behavior.

It should be stressed that the obtained microstructures are equivalent to trabecular bone architecture in terms of its mechanical properties and not on its geometrical features. However, having in mind the purpose of designing scaffold structures for bone tissue regeneration, the resulting microstructures designs may give an important insight into the fine geometry of scaffolds taking into account further design constraints ensuring, for example, appropriate mass transport properties. In the context of tissue engineering the designed bone substitutes would be customized for each anatomic region matching desirable local mechanical properties computed through the hierarchical bone remodelling model presented. Next section addresses the issue of scaffolds design taking into account local mechanical environment and mass transport properties.

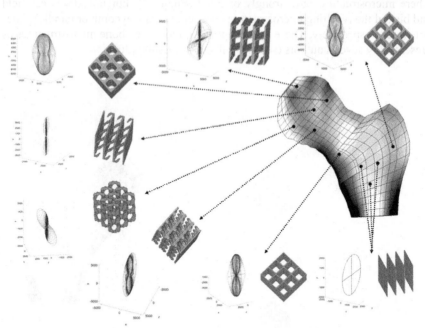

Fig. 7: Microstructures for trabecular bone. The bone regions presented in this figure show microstructures that are much more aligned in some spatial directions which accounts for an anisotropic behavior.

5 Scaffolds design

In this section some preliminary scaffold designs are obtained solving problem (9) combined with a local design requirement to impose a minimum bound for microstructure permeability. The scaffold is then conceived here as a periodic porous structure and designed for stiffness (maximized), volume fraction (upper bound constraint) and permeability (lower bound constraint).

The measure of permeability is obtained by homogenizing a potential flow problem in periodical porous media characterized by the Darcy Law (Darcy flow is assumed). The homogenization technique is the same used to derive the homogenized elasticity tensor presented before in this work. The resulting homogenized permeability tensor is enforced to be diagonal ($K^H_{ij} = 0$, $i \neq j$) and each diagonal component must be greater than an established limit ($K^H_{ij} \geq p_{min}$, $i = j$) as it accounts for minimum permeability in all spatial directions (orthotropic permeability tensor). The limit is set between 0 and 1 (or 0% and 100%) since permeability here is seen as a relative quantity. If the base cell domain is full then the permeability is 0 (minimum). If it is void then permeability is 1 (maximum).

Since problem (9) deals with the maximization of the strain energy density for a given local strain field and the extra constraint enforces orthotropic permeability, the resulting design solutions combine mechanical efficiency with desirable mass transport properties vital for the biological success of the scaffold.

Three particular mechanical environments or load conditions are considered in the examples shown in this preliminary study: hydrostatic load case (figure 8a), shear loads in two orthogonal planes (figure 8b) and multiload case (figure 8c). This multiload case combines the load case in figure 8b with two normal loads indicated in figure 8c which represent a maximum shear stress state at a plane inclined 45°. Each load case in figure 8c is weighted equally, i.e., $\alpha = 0.5$.

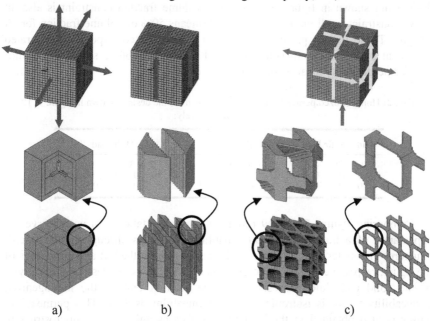

a)   b)   c)

Fig. 8: Microstructure designs with maximized stiffness for different local strain fields. Top shows local strain/stress field applied. Center shows a single cell design. Bottom shows periodical pattern design. a) hydrostatic load case; b) shear loads in two orthogonal planes, c) multiload case.

The results shown in figure 8 correspond to the maximization of the strain energy density subjected to the local volume fraction constraint only (maximum volume fraction is set to 50%). Figure 8c shows also two different perspectives of the resulting microstructure design, oblique and isometric views, respectively. As clearly seen the stiffest design solutions present closed wall cells or at least one orientated closed wall preventing permeability at all in the normal spatial direction [31]. The homogenized permeability tensors calculated for each design in figure 8 are shown in table 1. These results lead to the conclusion that mechanical efficiency works against the biological requirement that the microstructure must be porous and permeable.

Tab. 1: Homogenized permeability tensors obtained for the designs shown in figure 8.

Hidrostatic load	Shear load	Multiload
$\begin{bmatrix} 0 & 0 & 0 \\ 0 & 0 & 0 \\ 0 & 0 & 0 \end{bmatrix}$	$\begin{bmatrix} 0.223 & 0 & 0.223 \\ 0 & 0.446 & 0 \\ 0.223 & 0 & 0.223 \end{bmatrix}$	$\begin{bmatrix} 0.144 & 0.144 & 0.144 \\ 0.144 & 0.144 & 0.144 \\ 0.144 & 0.144 & 0.144 \end{bmatrix}$

In order to achieve permeable solutions an orthotropic permeability constraint is added to the stiffness maximization problem such that a minimum permeability of 0.3 (30%) is assured in all spatial directions. The resulting microstructure designs are shown in figures 9 to 11. The volume fraction constraint is also an active constraint in all these designs, which means 50% of volume fraction for all designs. These figures present the single cell and the corresponding generated periodical pattern for both solid and fluid domains. Table 2 shows the homogenized permeability tensors for the optimal designs.

Tab. 2: Homogenized permeability tensors obtained for the designs shown in figure 9, 10 and 11, respectively.

Hidrostatic load	Shear load	Multiload
$\begin{bmatrix} 0.3 & 0 & 0 \\ 0 & 0.3 & 0 \\ 0 & 0 & 0.3 \end{bmatrix}$	$\begin{bmatrix} 0.3 & 0 & 0 \\ 0 & 0.3 & 0 \\ 0 & 0 & 0.3 \end{bmatrix}$	$\begin{bmatrix} 0.3 & 0 & 0 \\ 0 & 0.3 & 0 \\ 0 & 0 & 0.3 \end{bmatrix}$

Comparing figures 8a and 9 (both for the hydrostatic case) we observe the appearance of a hole in the walls enabling fluid flow throughout the microstructure. The material and fluid domains are shown on the left and right sides of the figure, respectively. This representation clearly shows the desired connectivity of the solid and fluid phases. As can be seen in table 2 the homogenized permeability tensor is isotropic and the permeability is 30%. This permeability tensor is also obtained in the next two load cases although the microstructure design differs substantially. The results obtained for shear and multiload situations are shown in figures 10 and 11.

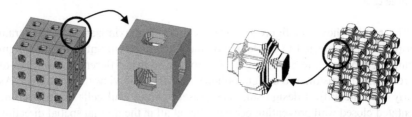

Fig. 9: Hydrostatic load case. Design with maximized stiffness and orthotropic permeability (minimum permeability is set to 30%). Left side is the material domain. Right side is the fluid domain.

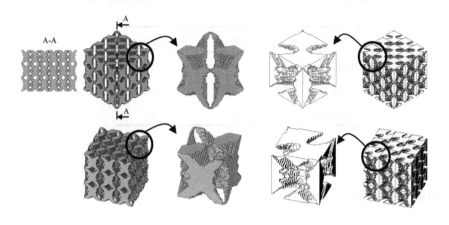

Fig. 10: Shear loads in two orthogonal planes. Design with maximized stiffness and orthotropic permeability (minimum permeability is set to 30%). Left side is the material domain. Right side is the fluid domain.

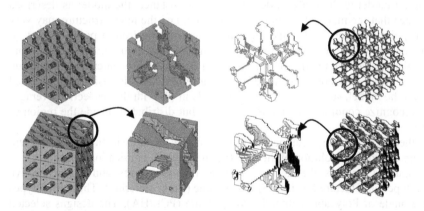

Fig. 11: Multiload case. Design with maximized stiffness and orthotropic permeability (minimum permeability is set to 30%). Left side is the material domain. Right side is the fluid domain.

As the stiffest designs presented in figure 8 are modified to accommodate the imposed permeability constraint (figures 9-11), some variations on microstructure stiffness and surface area take place (see table 3). The surface area measurement used here is the same presented in [16].

Tab. 3: Microstructure stiffness and surface area variation as a consequence of the applied permeability design constraint.

Load case	Stiffness	Surface area
Hydrostatic load	−33%	−7%
Shear load	−24%	+87%
Multiload	−28%	+15%

6 Microstructures Fabrication

It is our claim that the model presented can, with success, assist in the design of scaffold microstructures. As previously observed the microstructures obtained with the hierarchical remodelling model are mechanically equivalent to trabecular bone and well adapted to the respective mechanical environment . Additionally the imposition of local constraints, e.g. minimum permeability, facilitates the proper biological environment to promote bone growth.

The challenge now is the fabrication of such scaffold solutions.

There are some practical issues regarding the application of the current material model to the scaffold design field. For instance, the model as described assumes that the material is point wise periodic, i.e., the microstructure may vary from point to point. Truly this might be a shortcoming from a point of view of manufacturability. However, it is easy to impose, within the present model that the microstructure remains constant within a bone region of interest (e.g. the sub domain where the bone substitute is to be inserted). This is easily accomplished within the model, substituting the strain fields in (9) with the respective averages and recognize condition (9) not as pointwise, but distributed inside the region of interest.

Another aspect is the assurance of minimum size features in the microstructure design due to the resolution of the SFF (Solid Freeform Fabrication) machines.

Figure 12 shows the design of some microstructures and the respective prototype manufactured using a SFF technique (Laser sintering). These prototypes were made of Polycaprolactone/Hydroxyapatite (PCL/HA). The designs selected were collected as local solutions of the hierarchical bone remodeling model with a local permeability design constraint. These results demonstrate the feasibility of produce scaffolds tailored for a given bone site.

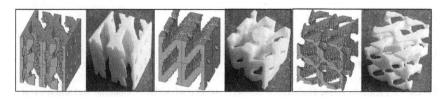

Fig. 12: Microstructure design models (periodic patterns) and respective prototype fabricated by a SFF technique (These prototypes were fabricated using the facilities of the Scaffold Tissue Engineering Group at the University of Michigan).

7 Concluding Remarks

In this chapter a three-dimensional multiscale model to simulate the functional adaptation of bone was presented. The model characterizes the morphology of bone on a macroscopic level (apparent density) and simultaneously characterizes the local microstructure of trabecular bone with the respective anisotropic properties of bone. The multiscale law of bone remodelling is derived based on a mechanical stimulus to produce stiff bone regulated by a biologic parameter that controls the amount of bone apposition/resorption.

Results show that the model reasonably reproduces the functional gradation of natural bone, however it was noted that the two criteria used to derive the present law of bone remodelling are not enough to reproduce the actual architecture of bone and one needs to include, within the model, additional biologic requirements related to factors such as permeability and vascularisation. Furthermore, the resulting microstructures should be understood as equivalent to trabecular architecture in terms of mechanical behaviour and not in terms of its geometry. The microstructures obtained by the multiscale model characterize the bone behaviour and the model is suitable to assist the design bone substitutes used for tissue regeneration. Having this application in view, some microstructure prototypes were built using SFF technique showing the feasibility of this approach for scaffold design.

Acknowledgements

This work was supported by FCT through the Projects PTDC/EME-PME/71436/ 2006, PTDC/EME-PME/104498/2008 and PhD scholarship – SFRH/BD/25033/ 2005 (P. Coelho). The results presented here were produced using the IST Cluster (IST/Portugal). We would like to thank to Prof. Scott Hollister (University of Michigan) for the microstructures prototypes.

References

1. Pompe W, Worch H, Epple M, Friess W, Gelinsky M, Greil P, Hempel U, Scharnweber D, Schulte K (2003), Functionally graded materials for biomedical applications, Mat Sci Eng A362:40-60.
2. Lin CY, Kikuchi N and Hollister SJ (2004) A novel method for biomaterial scaffold internal architecture design to match bone elastic properties with desired porosity. J Biomech, 37:623-636.
3. Wolff J, (1986) The law of bone remodeling (Das Gesetz der Transformation der Knochen, Hirschwald, 1892). Translated by Maquet P. and Furlong R. Springer, Berlin.
4. Hart RT, Davy DT, Heiple KG (1984). A computational model for stress analysis of adaptive elastic materials with a view toward applications in strain-induced bone remodeling. J Biomech Eng 106:342-350.
5. Huiskes R, Weinans H, Grootenboer HJ, Dalstra M, Fudala B, Sloof TJ (1987) Adaptive bone-remodeling theory applied to prosthetic-design analysis. J Biomech 20:1135-1150.
6. Carter DR, Fyhrie DP, Whalen RT (1987) Trabecular bone density and loading history: regulation of tissue biology by mechanical energy. J Biomech 20:785-795.
7. Beaupré GS, Orr TE, Carter DR, (1990) An approach for time-dependent bone modeling and remodeling-theoretical development. J. Orth Res 8:651-661.
8. Weinans H, Huiskes R, Grootenboer HJ (1992) The behavior of adaptive bone- remodeling simulation models. J Biomech 25:1425-1441.
9. Cowin SC, Sadegh AM, Luo GM (1992) An evolutionary Wolff's law for trabecular architecture, J Biomech Eng 114:129-137.
10. Prendergast PJ, Taylor D (1994). Prediction of bone adaptation using damage accumulation. J Biomech 27:1067-1076.
11. Hart RT, Fritton SP (1997) Introduction to finite element based simulation of functional adaptation of cancellous bone. Forma 12:277-299.
12. Jacobs CR, Simo JC, Beaupré GS, Carter DR (1997) Adaptive bone remodeling incorporating simultaneous density and anisotropy considerations. J Biomech 30:603-613.
13. Fernandes P, Rodrigues H, Jacobs C (1999) A model of bone adaptation using a global optimization criterion based on the trajectorial theory of Wolff. Comput Methods Biomech Biomed Eng 2:125-138.
14. Rodrigues H, Jacobs C, Guedes J, Bendsøe M (1999) Global and local material optimization applied to anisotropic bone adaptation. In: Perdersen P, Bendsoe MP (ed), Synthesis in Bio Solid Mechanics. Kluwer Academic Publishers.
15. Doblaré M, Garcia JM (2002) Anisotropic bone remodelling model based on a continuum damage-repair theory. J Biomech 35:1-17.
16. Coelho PG, Fernandes PR, Rodrigues HC, Cardoso JB and Guedes JM (2009) Numerical modeling of bone tissue adaptation – A hierarchical approach for bone apparent density and trabecular structure, J. Biomech. 42:830-837.
17. Lucchinetti E (2001) Composite Model of Bone Properties. In: Cowin SC (ed) Bone Mechanics Handbook, 2nd Edition, CRC Press.
18. Guedes J, Kikuchi N, (1990) Preprocessing and postprocessing for materials based on the homogenisation method with adaptive finite element method. Comput Meth Appl Mech Eng 83:143-198.
19. Rodrigues H, Guedes J, Bendsøe M (2002) Hierarchical optimisation of material and structure. Int J Struct Multidisc Optim 24:1-10.
20. Bendsøe MP, Sigmund O, (2003) Topology optimization theory, methods and applications. Springer, Berlin Heidelberg New York.
21. Martin RB, Burr DB, Sharkey NA (1998) Skeletal Tissue Mechanics. Springer-Verlag, NY.
22. Svanberg K (1987) The method of moving asymptotes – a new method for structural optimisation. Int J Numer Methods Eng 24:359-373.

23. Fleury C (1989) CONLIN: an efficient dual optimizer based on convex approximation concepts. Struct Optim 1:81-89.
24. Coelho PG, Fernandes PR, Guedes JM, Rodrigues HC, 2008. A hierarchical model for concurrent material and topology optimization of three-dimensional structures. Struct Multidisc Optim 35:107-115.
25. Viceconti M, Casali M, Massari B, Cristofolini L, Bassini S, Toni A (1996) The 'Standardized Femur Program' Proposal for a Reference Geometry to be Used for The Creation of Finite Element Models of the Femur. J Biomech 29:1241.
26. Bergmann G (1998) Hip98 – Loading of the hip joint. Free University of Berlin. Compact disc, ISBN 3980784800
27. Bergmann G, Deuretzbacher G, Heller M, Graichen F, Rohlmann A, Strauss J, Duda G (2001) Hip contact forces and gait patterns from routine activities. J Biomech 34:859-871.
28. Coelho PG, Fernandes PR, Rodrigues HC (2010) Multiscale Modeling of Bone Tissue with Surface and Permeability Control, J. Biomech DOI:10.1016/j.jbiomech.2010.10.007.
29. Ottosen NS, Ristinmaa M (2005) The mechanics of constitutive modeling. Elsevier, Oxford.
30. Gibson L, Ashby M (1988) Cellular Solids, Structure and Properties. Pergamon Press, Oxford, England.
31. Sigmund, O. (1999), On the optimality of bone structure. Proc. of IUTAM Symposium on Synthesis in Bio Solid Mechanics, Copenhagen, Denmark, 221-233.

Nonlinear Elastic Scaffold Design, Modeling and Fabrication for Soft Tissue Engineering

Scott J. Hollister, Claire G. Jeong, J. Jakob Schwiedrzik, Anna G. Mitsak, Heesuk Kang, Francesco Migneco

Abstract Biologic soft tissues exhibit nonlinear elastic behavior under physiologic forces. It is widely postulated that within tissue engineering, biomaterial scaffolds should be designed to replicate native tissue behavior. For soft tissue engineering, this implies that scaffolds should be designed to exhibit nonlinear elasticity. However, unlike linear elasticity, there is no single constitutive model for nonlinear elasticity. In this chapter, we discuss candidate strain energy functions that can be used to both model soft tissue behavior and biodegradable elastomer behavior. We further demonstrate how designed pore architecture not only reduces the tangent stiffness of nonlinear elastic biomaterials, but decreases the nonlinearity of such materials in accordance with known upper bounds on effective nonlinear elastic properties. Finally, we demonstrate that nonlinear elastic finite element analysis on porous scaffold unit cell models can be used to predict the effective nonlinear elastic properties of elastomeric scaffolds.

1 Introduction

Scaffolds are a significant component of any tissue engineering approach, delivering biologics (cells, genes, and/or proteins) and interacting with host cells on a hierarchical scale. It is the scaffold interaction with the host cell that ultimately determines the success or failure of any tissue reconstruction. This

Scott J. Hollister
Scaffold Tissue Engineering Group, Department of Biomedical Engineering, The University of Michigan, Ann Arbor, MI USA
Department of Mechanical Engineering, The University of Michigan, Ann Arbor, MI USA
Department of Surgery, The University of Michigan, Ann Arbor, MI USA
email: scottho@umich.edu

Claire G. Jeong, J Jakob Schwiedrzik, Anna G. Mitsak, Heesuk Kang, Francesco Migneco
Scaffold Tissue Engineering Group, Department of Biomedical Engineering, The University of Michigan, Ann Arbor, MI USA

P.R. Fernandes and P.J. Bártolo (eds.), *Advances on Modeling in Tissue Engineering*,
Computational Methods in Applied Sciences 20, DOI 10.1007/978-94-007-1254-6_3,
© Springer Science+Business Media B.V. 2011

interaction, however, occurs through a number of paths, from the local cell material interaction (both attachment and cell deformation) occurring at the micron scale, to the nutrient/waste diffusion and deformation of millions of cells at the hundreds of microns to millimeter scale, to the whole tissue/organ scaffold interaction at the tens of centimeter scale. In short, we need to understand cell-material attachment, cell-material deformation mechanics, and cell-material mass transport characteristics to engineer a scaffold that reproducibly ensures optimal tissue regeneration. Such an intimidating task is further complicated by the fact that that these features, cell-material attachment, mechanics and mass transport, will differ between tissues and possibly between the same tissues at different anatomic locations.

The key question of course is: "How do we design a scaffold to meet all these requirements?" Before we can answer this question, an even more fundamental question is "How do we quantitatively characterize the scaffold design requirements?". There are two approaches to defining scaffold design requirements for enhanced tissue regeneration prior to experimentation. The first is based on assuming tissue response to local mechanical and mass transport stimuli (local strain, fluid flow, oxygen consumption, etc.). We denote this approach as the *tissue regeneration modeling* approach. In this approach, an assumption is made concerning how these local stimuli affect tissue regeneration (see for example, [1-3]). A scaffold design is proposed and tissue response is calculated. This approach allows parametric simulation of how scaffold design affects tissue regeneration. Although this approach is extremely useful for evaluating scaffold design and likely close to physiologic reality, it is not known how uncertainty in boundary condition and heterogeneous tissue constitutive properties will influence the how close the simulation matches real life. In this sense, such models must be continuously validated with experiments under controlled conditions.

The second approach relies on a basic postulate that existing native tissue mechanical and mass transport properties reflect a successful adaptation to the tissue *in situ* mechanical and mass transport environment. Thus, if we can engineer the scaffold to reflect the native tissue properties, the scaffold will provide a suitable averaged (since effective properties represent an average of local field variables like strain, fluid flow and oxygen consumption) environment for tissue regeneration. We denote this approach the *native tissue property* approach. This approach also has limitations, namely questions concerning what tissue properties and corresponding values of these properties make appropriate design values and the need to collect and test tissues from a large a sample population as possible.

Of course, the tissue regeneration modeling and native tissue property approaches for defining scaffold design targets are not mutually exclusive and developing improved scaffold designs is likely to require the use of both synergistically. As a starting point, we will focus on the native tissue property approach. In this sense, some papers have been published proposing design requirements for tissue engineering scaffolds [4-8], specifically proposing effective Young's modulus and permeability as design targets. However, despite these proposed design targets, we are no closer to knowing the optimal scaffold design, even for bone regeneration. This is because to achieve these design targets, we must first know mathematically the relationship between pore architecture, material and effective Hooke's constants and permeability. We must then be able to design architecture to attain desired effective constants, and then reproducibly fabricate these scaffolds with the designed architectures that can be tested within *in vivo* models.

It is even more humbling to realize that these proposed design targets of Hooke's elastic constants and constant permeability are likely only relevant for bone tissue reconstruction. Soft tissues exhibit more complex mechanical and mass transport behavior than hard tissues. Soft tissues are generally nonlinear elastic and viscoelastic, with mass transport characteristics including permeability and diffusion that are deformation dependent. Unlike linear elastic materials whose mechanical behavior is uniquely defined by Hooke's law, there is no unique description of nonlinear elastic or viscoelastic behavior. Since these models have often been developed to best fit a specific tissue behavior, it stands to reason that many models would exist. However, for purposes of scaffold design, creating standard models is important for measuring scaffold design mechanical behavior versus natural tissue behavior. Regarding deformation dependent mass transport, the issue is not an abundance of models, but rather a dearth. In this case, there may not be a model by which to measure scaffold design.

Thus, to begin developing scaffolds for functional soft tissue engineering, we must first develop design targets based on natural soft tissue constitutive models. From here, we must be able to design the pore architecture of scaffolds made of known nonlinear elastic material and predict their effective nonlinear elastic and deformation dependent mass transport properties. These designs must be fabricated reproducibly and their mechanical and mass transport properties measured and compared with the design targets of natural tissue effective properties. Finally, once characterized viz a viz natural tissue properties, the scaffolds must be experimentally tested for soft tissue regeneration. Once the results are known, we can begin to test specific design hypotheses as to how scaffold effective nonlinear elastic properties affect soft tissue regeneration.

2 Standardized Nonlinear Elastic Models for Scaffold Design Targets

Soft tissues from a wide variety of anatomic locations exhibit a nonlinear stress strain relationship with hysteresis. Given that the soft tissue hysteresis becomes small and reproducible after cyclic loading, many investigators have adopted the postulate first put forth by YC Fung [9-11] that soft tissues could be accurately be modeled as "pseudoelastic" materials with a nonlinear elastic strain energy function ignoring the viscoelastic effects (see also [12]). Once the pseudoelastic principal is established, we then can apply a wide range of hyperelastic strain energy functions [12, 13] to model soft tissue nonlinear elasticity. For characterization of tissue mechanics, a wide variety of strain energy functions have been proposed. However, for design purposes it is good to standardize the use of a limited number of strain energy functions for the purpose to defining natural tissue properties as design targets and fitting effective scaffold properties for comparison with the natural tissue properties. In this chapter we propose the use of a three strain energy functions widely used in tissue mechanics as standards for nonlinear elastic scaffold design targets.

2.1 Fung 1D Exponential Strain Energy Function

A widely used and relatively simple strain energy function was first proposed by YC Fung in 1967 [9]. Fung observed that many tissues displayed a classic stiffening response with increasing deformation, and further that the tangent modulus was linearly correlated to the 1st Piola-Kirchoff stress T by:

$$\frac{dT}{dE} = B(T + A) \tag{1}$$

where E is the large strain and B and A are empirical constants in the model. Integrating eq. 1 and using the fact that the stress is zero when the strain is zero yields the stress-strain relationship:

$$T = Ae^{BE} - A \tag{2}$$

with the associated strain energy function:

$$W = \frac{A}{B}e^{BE} - AE \tag{3}$$

This simple model offers the advantage that is simple and the constants A and B are readily interpreted for design purposes. The constant B characterizes the degree of effective nonlinearity, with higher values indicating more highly nonlinear stress strain relationships (Fig 1).

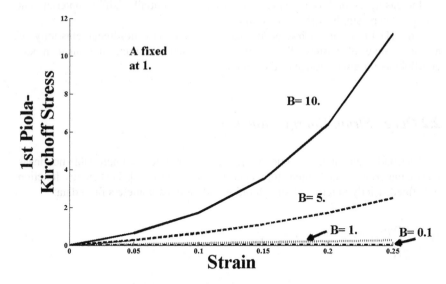

Fig. 1. Increasing parameter B increases stress strain nonlinearity in Fung model

The constant A influences the initial tangent modulus of the nonlinear material, with increasing A producing a stiffer material (Fig. 2).

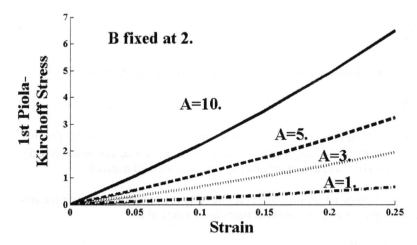

Fig. 2. Increasing parameter B increases initial stiffness for fixed B but does not significantly affect the nonlinearity of the stress strain curve.

Increasing A and B together will produce an initially stiffer material that becomes increasingly stiffer with strain.

The 1D Fung model has been used to characterize nonlinear elasticity of a wide range of tissues. We recently published 1D Fung nonlinear model coefficients for a wide range of tissues [6].

2.2 Ogden Strain Energy Function

A second strain energy function proposed for isotropic, incompressible nonlinear elastic materials is the Ogden strain energy function [13, 14]. This energy function is defined strictly in terms of stretch ratios and model parameters fit to data:

$$W = \sum_{n=1}^{N} \frac{\mu_n}{\alpha_n} \left(\lambda_1^{\alpha_n} + \lambda_2^{\alpha_n} + \lambda_3^{\alpha_n} - 3 \right) \tag{4}$$

where λ_1, λ_2, and λ_3 are the stretch ratios in the 1,2, and 3 directions, respectively, and μ_n and α_n are model parameters fit to data. An important characteristic is physical interpretation and limits on the model parameters. The sum of the μ_n and α_n products is the shear modulus μ in linear elasticity [13, 14]:

$$\mu = \sum_{n=1}^{N} \mu_n \alpha_n \tag{5}$$

which requires that each product in the summation be greater than zero:

$$\mu_n \alpha_n > 0 \tag{6}$$

Equation 6 ensures that if μ_n is chosen to be positive, α_n must be positive, which is important for design purposes since both μ_n and α_n can monotonically increase between 0 and an upper bound for the base scaffold material.

The 1st Piola-Kirchoff (PK) stress is related to the principal stress components for an isotropic, incompressible nonlinear elastic material by:

$$T_{ii} = -p \frac{1}{\lambda_i} + \frac{\partial W}{\partial \lambda_i} \ ; \ no \ sum \ on \ i \ for \ T \tag{7}$$

where p is the hydrostatic pressure. We use the 1st PK stress since this is commonly determined in soft tissue experiments. To fit the Ogden model to soft tissue behavior, we typically perform a uniaxial or biaxial test with at least one face is traction free ($T_{ij} = 0$). This allows us to solve for p in and use p in calculation of stress in the loaded directions. If we take x_3 as the loaded direction in a uniaxial test, then the expression for the 1st PK stress becomes:

$$T_{33} = -\frac{\lambda_1}{\lambda_3} \sum_{n=1}^{N} \mu_n \lambda_1^{\alpha_n - 1} + \sum_{n=1}^{N} \mu_n \lambda_3^{\alpha_n - 1} ;$$

(8)

$$N = 1: T_{33} = -\frac{\lambda_1}{\lambda_3} \mu_1 \lambda_1^{\alpha_1 - 1} + \mu_1 \lambda_3^{\alpha_1 - 1}$$

To better understand the behavior of the Ogden model, consider the influence of both μ and α for the 1-term model ($N=1$ in eq. 4). Increasing α from 0.1 to 5 (with μ fixed at 1.0) increases stress-strain nonlinearity (Figure 3) with strain stiffening in compression and strain softening in tension (Fig. 3).

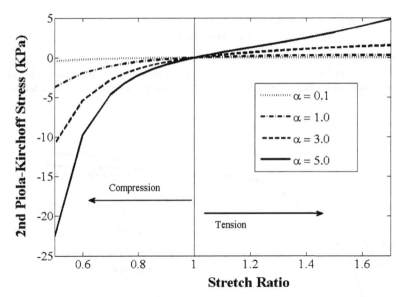

Fig. 3. Nonlinear compression tension behavior of the Ogden model for the model parameter α ranging from 0.1 to 5. Strain stiffening occurs in compression while strain softening occurs in tension. Decreasing α results in more linear behavior.

As the parameter α increases past 5, there is increasing strain stiffening in compression while the tensile behavior switches from strain softening to strain stiffening (Fig. 4).

As the parameter μ increases it also increases the nonlinearity in both tension and compression (Fig. 5). However, if α is maintained constant as μ changes, there is no transition from strain softening to strain stiffening in tension.

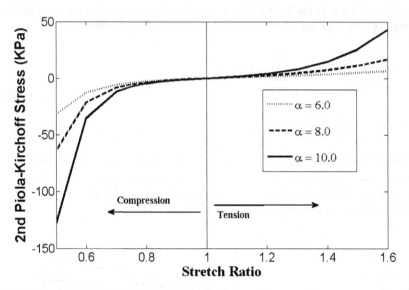

Fig. 4. Nonlinear compression tension behavior of the Ogden model for the model parameter α ranging from 6. to 10. Strain stiffening continues to occur in compression while tensile behavior transitions from strain softening to strain stiffening. Decreasing α again results in more linear behavior.

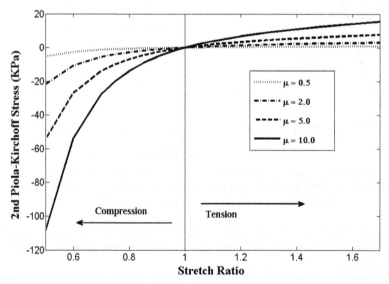

Fig. 5. Nonlinear compression tension behavior of the Ogden model for the model parameter μ ranging from 0.5 to 10. Strain stiffening continues to occur in compression while tensile behavior exhibits strain softening to strain stiffening. Decreasing μ results in more linear behavior.

2.3 Anisotropic Extension of the Ogden Strain Energy Function for Fiber Reinforced Tissues

As noted by Fung [9], most soft tissues exhibit strain stiffening behavior even in tension. Many researchers postulate that this strain stiffening behavior is due embedded fibers that are initially crimped but uncrimp during stretching, becoming stiffer. In this case, would one expect stiffening behavior only along the fiber axis. Holzapfel and Ogden [15-17], as well as Humphrey et al. [18-19] among others have proposed many models for fiber-reinforced biologic tissues in which a ground substance is assumed to be isotropic and anisotropic stiffening behavior is assumed to be an additive or multiplicative addition to the isotropic ground substance. In general, the strain energy function then takes the form

$$W = W^{isotropic} + W^{anisotropic} + W^{interaction} \qquad (9)$$

where the term $W^{interaction}$ represents the interaction between the ground matrix and the fibers. To simplify the approach for design purposes, we will consider the anisotropic strain energy function to be a sum of isotropic and anisotropic components:

$$W = W^{isotropic} + W^{anisotropic} \qquad (10)$$

The contribution of the deforming fibers to the total strain energy in two orthogonal directions is characterized using the following invariants:

$$I^4 = a_i^0 C_{ij} a_j^0; \; I^6 = b_i^0 C_{ij} b_j^0 \qquad (11)$$

where a and b are vectors describing the two different fiber orientations in the reference configuration, C_{ij} is the right Cauchy deformation tensor, and I^4 and I^6 are the invariants. Holzapfel et al. [15, 16] proposed a fiber extension stiffening anisotropic component that can account for both transversely isotropic (if one fiber family is included with invariant I^4) and orthotropic (if two fiber families are include using a second invariant I^6):

$$W^{aniso} = \frac{k_1}{2k_2} \left\{ e^{\left[k_2 (I^4 - 1)^2 \right]} - 1 \right\} + \frac{k_1}{2k_2} \left\{ e^{\left[k_2 (I^6 - 1)^2 \right]} - 1 \right\} \qquad (12)$$

where I^4 and I^6 are invariants that describe fiber deformation, and k_1 and k_2 are model parameters. Holzapfel et al. [16] note that k_1 and k_2 should be positive, which is important for design purposes as they will then exhibit a monotonic behavior from 0 to the properties of a base scaffold material. One issue for the anisotropic component in eq. 10 is that it is postulated that fibers carry load only when extended, which means only when I^4 and I^6 are greater than 1 (I^4 and $I^6 = 1$ is the zero deformation case). Thus, the we propose a slight modification of the original Holzapfel anisotropic strain energy function to use the general Ogden strain energy function (eq. 4) for the isotropic component:

$$W = \overbrace{\sum_{n=1}^{N} \frac{\mu_n}{\alpha_n} \left(\lambda_1^{\alpha_n} + \lambda_2^{\alpha_n} + \lambda_3^{\alpha_n} - 3\right)}^{Isotropic\ Component} +$$

$$\underbrace{\frac{k_1}{2k_2}\left\{e^{\left[k_2(I^4-1)^2\right]} - 1\right\} + \frac{k_1}{2k_2}\left\{e^{\left[k_2(I^6-1)^2\right]} - 1\right\}}_{Anisotropic\ Component-active\ during\ fiber\ extension}$$

(13)

If we then apply eq. 13 in eq. 7 and assume the fibers are oriented in the x_1 and x_3 direction, we obtain the following expression for the 1st PK stress component T_{33}:

$$T_{33} = -\frac{\lambda_1}{\lambda_3} \sum_{n=1}^{N} \mu_n \lambda_1^{\alpha_n-1} + \sum_{n=1}^{N} \mu_n \lambda_3^{\alpha_n-1} +$$

$$\left\{ \begin{array}{l} -2k_1 \frac{\lambda_1^2}{\lambda_3} a_1^0 a_1^0 (I^4 - 1)e^{\left[k_2(I^4-1)^2\right]} + \\[2ex] 2\, b_3^0 b_3^0 \lambda_3 k_1 (I^6 - 1)e^{\left[k_2(I^6-1)^2\right]} \end{array} \right\}$$

(14)

3 Porosity Effects on Nonlinear Elasticity: Experimental and Computational Studies on Elastomeric Scaffolds

Fitting nonlinear elastic models such as those in section 2 to soft tissue experiments provides targets for functional scaffold design. Thus, if we are seeking to design functional aspects of soft tissue scaffolds, fits of nonlinear constitutive models to soft tissue mechanical behavior provide quantitative targets we can seek to match with biomaterial scaffolds. Scaffolds of course must be porous to deliver biologics like cells, genes and proteins. Therefore, our ability to match tissue nonlinear elasticity depends on 1) the constitutive properties of the base scaffold material, and 2) the influence of the pore architecture we introduce into the base scaffold material. Our design criteria for scaffold function is to match the effective scaffold constitutive properties to tissue nonlinear elastic properties. If the base scaffold material is linear, then we must introduce geometric nonlinearity such as contact, buckling, etc. to create effective nonlinear elastic behavior. For this chapter, we will instead focus on base scaffold materials that exhibit nonlinear elastic behavior and how the introduction of porosity alters these base nonlinear elastic properties.

An important starting question is whether we can estimate *a priori* how close a scaffold fabricated from a nonlinear biomaterial can match the nonlinear behavior of a given soft tissue. In this regard, bounds on effective anisotropic nonlinear elastic behavior would be extremely useful. However, bounds for the anisotropic behavior of even linear elastic microstructured materials are few and often complex to implement [20]. As one might expect, bounds on the effective behavior of nonlinear elastic materials are even more scarce. Indeed, only the most general Voight and Reuss type bounds on the behavior of general anisotropic nonlinear elastic materials, as described by Ogden [21]. The upper Voight type bounds posit that the effective strain energy function of a nonlinear microstructure material is equal to the strain energy function of each phase multiplied by the volume fraction of that phase:

$$W^{effective} = \sum_{n=1}^{N} v_f^n W^n \tag{15}$$

the complementary lower Reuss type bounds posits the same relationship, only for the complementary stress energy functions. The difficulty for porous materials like scaffolds is that the pores have zero strain energy and therefore infinite stress energy. This means that the Reuss bound will produce zero strain energy as the lower bound for all porous materials. Such a material can of course be theoretically contemplated as isolated material particles separated by space. Tighter bounds have been derived for isotropic nonlinear elastic materials, but only for specialized isochoric deformations (volume preserving) in 3D [22] and only under general deformations for Neo-Hookean nonlinear elastic materials in 2D [23]. Thus, the absolute lower bound for any linear or nonlinear porous material can always be shown to be zero. The most general upper bound for a nonlinear elastic material can be shown to be the base material strain energy functions linear weighted by the volume fraction of each phase. Still, this upper bound does provide information on how to design a porous material with a given volume fraction in comparison to a given soft tissue nonlinear constitutive model target.

To better approximate soft tissue compliance and nonlinear behavior, a new class of bioresorbable elastomeric polymers have been introduced for soft tissue engineering. Yang et al. [24] introduced Poly(Octanediol-co-Citrate) (POC) as a bioresorbable elastomer for soft tissue engineering. Wang et al. [25] introduced Poly(Glycerol-co-Sebacate) (PGS) as a bioresorbable elastomer for soft tissue engineering. Our own group has developed new methods to create designed scaffold architectures from both POC [26, 27] and PGS [27, 28]. Finally, our own group [29] has introduced Poly Dodeacanoic Acid (PGD) as a new bioresorbable elastomer exhibiting temperature dependent and shape memory features for soft tissue engineering. All three of these bioelastomers exhibit nonlinear elastic behavior as illustrated by stress-strain curves (although PGD only exhibits such behavior over 37°C, close to body temperature). However, explicit fit of nonlinear elastic constitutive models to this class of biodegradable elastomers was first performed by our group [26, 28, 29].

Specifically, we fabricated solid dogbone specimens according to ASTM 412A of all three bioelastomers (POC, PGS, PGD; Fig. 6).

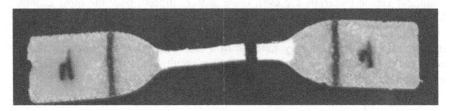

Fig. 6. Example dog bone tensile specimen after test to failure. These types of specimens were used to test mechanical properties of base elastomeric biomaterials.

We furthermore varied synthesis conditions as this varies material characteristics like cross-linking density and thus the nonlinear elastic behavior. After fabrication, specimens were tested in tension at a crosshead speed of 1mm/min. Typical stress-strain curves demonstrate classic rubber-like nonlinear elasticity (note PGD was tested in a controlled chamber at 37°C). We measured volume changes at different strain levels and found approximately 3% volume change, thus validating the incompressibility assumption. Following testing, we fit a 1-term isotropic Ogden model (eq. 8) to all solid specimens results. Solid base POC, PGS, and PGD nonlinear behaviors were all fit well (R^2 for all fits > 0.95) by the Ogden model (Table 1).

Table 1. Base nonlinear elastic properties fit to a 1-term Ogden model for bioresorbable elastomers POC, PGS, PGD. Model parameters depend on synthesis conditions including curing temperature and time, which alters crosslink density.

Material	Synthesis Conditions	μ_1 Ogden Model (KPa)	α_1 Ogden Model
POC	100°C; 5 days	202 ± 23	1.62 ± 0.18
POC	100°C; 4 days; 120°C 1 day	194 ± 15	1.86 ± 0.21
PGS	120°C; 2 days	172 ± 20	2.13 ± 0.07
PGS	135°C; 2 days	100 ± 17	1.98 ± 0.02
PGS	135°C; 3 days	291 ± 22	1.91 ± 0.03
PGD	90°C; 2 days; 120 C 2 days	300 ± 12	2.14 ± 0.5

Our group has also been able to fabricate porous scaffolds from POC and PGS, using designed solid free-form fabricated techniques (Figure 7). These designed porous scaffolds had either cylindrical pores of 32% (c32), 44% (c44), or 62% (c62) porosity or a spherical pore with 50% porosity (s50).

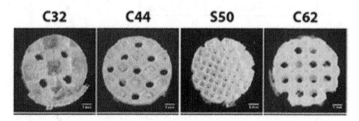

Fig. 7. Example of bioresorbable elastomers with designed pore structure. From left to right: C32 cylindrical pores with 32% porosity, C44 cylindrical pores with 44% porosity, S50 spherical pores with 50% porosity, and C62 cylindrical pores with 62% porosity.

We fit a Neo-Hookean nonlinear elastic model (The Ogden model of eq. 4 with one term and the α exponent fixed at 2) to both solid POC as well as the designed POC and PGS scaffolds in compression. Results showed that the Neo-Hookean model fit the nonlinear elastic behavior of all designs well (all $R^2 > 0.95$, Table 2).

Table 2. Fit of Neo-Hookean model to compressive solid and scaffold POC tests. Comparison with Voight upper bound (eq. 15).

Design/Material	Upper Bound	$\mu 1$ Neo-Hookean Model (KPa)	R^2 Fit
POC Solid	----------------	322 ± 71	0.97 ± 0.01
POC c32	219	185 ± 18	0.97 ± 0.01
POC c44	180	88 ± 7	0.99 ± 0.01
POC c62	122	35 ± 12	0.98 ± 0.01
POC s50	161	64 ± 2	0.99 ± 0.01
PGS s50	161	57 ± 13	0.98 ± 0.01

The POC results also fell within the upper Voight bounds proposed for micro-structured nonlinear elastic materials by Ogden [21]. The results of increasing scaffold porosity experimentally caused a decrease in both the initial tangent modulus as well as the degree of nonlinearity (Figure 8).

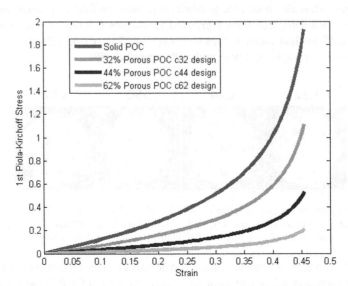

Fig. 8. Average Neo-Hookean fit results for solid an porous scaffold POC demonstrate decreased stiffness and nonlinearity with increasing porosity from Solid POC (red) to 62% porous cylindrical pore design scaffold (cyan).

It is interesting that introducing porosity into a material whose base (solid) properties are nonlinear elastic decreases not only the stiffness, as one sees with a linear elastic material, but also dramatically decreases the degree of nonlinearity. This is likely due to the loss of material in the pores that decreases the amount of stiffening with deformation. In fact, this loss of nonlinearity is predicted by the Voight type bounds for porous nonlinear microstructured materials proposed by Ogden (eq. 15).

To determine how closely we could predict the effective nonlinear elastic POC scaffold properties, we created nonlinear elastic finite element models of the scaffold pore design and subjected these to similar compressive loading as in the experiment. The commercial finite element code Abaqus was used to analyze the geometrically and material nonlinear problems. The base coefficients for solid POC were input using the Neo-Hookean parameters for POC under the second set of curing conditions. We utilized both a voxel and smooth tetrahedral finite element models of the basic unit cell (Figure 9).

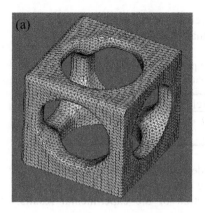
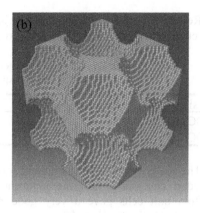

Fig. 9. Examples of both (a) smooth tetrahedral and (b) voxel finite element models of designed unit cells used for the geometric and material nonlinear analysis.

For the voxel models, the one face was deformed in the z direction while the other face was constrained in the z direction but allowed to slide in the x and y directions. For the tetrahedral model, the model had additional constraints in that the faces were allowed to slide but not deform outward. The stretch ratios were computed from the applied displacements and the effective 1st Piola-Kirchoff z stress component was calculated from the total reaction force on the constrained face divided by the initial area of the unit cell. The effective Neo-Hookean constants were calculated using the same curve fitting techniques as for the experimental results. The finite element analyses demonstrated the propensity of the porous scaffolds to exhibit heterogeneous stress distributions as well as instabilities in thin scaffold walls under large strain (Fig. 10).

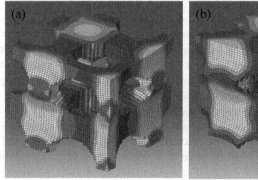
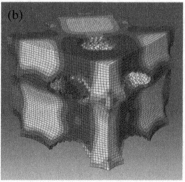

66% Porous Cylindrical Pore   53% Porous Spherical Pore

Fig. 10. Example stress contour and deformation of scaffold unit cells under uniform compression. Note the large deformation in thin scaffold walls for both (a) 66% cylindrical porous design and (b) 53% spherical pore design.

Results from the voxel models matched well to the experimental data, while the tetrahedral model produced a stiffer result than the experiment (Table 3).

Table 3. Comparison of nonlinear finite element predictions of effective Neo-Hookean constants to experimental data.

Design	Experimental (KPa)	Voxel FE (KPa)	Smooth FE (KPa)
POC c62	35 ± 12	33.9	72.5
POC s50	64 ± 2	57.3	-------------------

The stiffer results from the tetrahedral model was likely due to the additional boundary constraints. Still, the nonlinear finite element analyses of the designed scaffold unit cells gave a very good prediction of the experimental results, especially for the voxel model.

4 Summary

In summary, many investigators have proposed that biomaterial scaffold mechanical properties should match the mechanical properties of the tissue being replaced. Matching native soft tissue mechanical properties becomes more challenging due to the nonlinear elastic and anisotropic behavior of these tissues. This challenge becomes more apparent given that many widely used degradable biomaterials themselves do not exhibit nonlinear elastic properties. Finally, unlike linear elasticity which characterizes mechanical behavior using the unique Hooke's Law model, there is no unique constitutive model to characterize nonlinear elastic behavior. Indeed, a wide variety of isotropic and anisotropic strain energy functions have been proposed to characterize the nonlinear behavior of soft tissues.

While previous tissue strain energy functions have been proposed to capture the wide range of tissue behavior, the existence of multiple strain energy functions makes it difficult to develop consistent scaffold mechanical design targets. Therefore, we have proposed a modification of a commonly used anisotropic strain energy function proposed by Holzapfel that incorporates the widely used Ogden strain energy function with the anisotropic components given by Holzapfel as a design target for soft tissue nonlinear elastic behavior.

The ability to match nonlinear elastic soft tissue properties of course depends on the base material properties of the chosen biomaterial and how the material and voids are arranged in 3D space. If the material is linear elastic, effective nonlinear elasticity can only be created through microstructure geometric nonlinearities including contact and buckling within the porous microstructure. If the base material is nonlinear elastic, of course, effective nonlinear elastic behavior can be created through any design. We further demonstrated that recently developed bioresorbable elastomers (POC, PGS, and PGD) demonstrate both base nonlinear elastic behavior and designed effective nonlinear elastic behavior. Both solid and porous scaffold nonlinear elastic behavior can be fit well using both a 1-term Ogden strain energy function and a Neo-Hookean strain energy function.

Of course, we would like to be able to predict architecture design influences on effective scaffold nonlinear elastic behavior. In this sense, bounds on effective properties for a given amount of material are useful in that the define what effective properties are achievable for a given amount of material. However, only general Voight and Reuss type bounds are currently available for general strain energy functions under general deformation. The upper Voight bound gives a loose estimate of upper achievable nonlinear elastic properties, However, the lower Reuss bound for porous materials like scaffolds is zero. The use of nonlinear finite element models of porous architecture can provide good estimates of effective nonlinear elastic properties.

Looking forward, it is important to consider the anisotropic nature of soft tissue nonlinearity as currently modeled and the nonlinearity achievable with biomaterial scaffolds. Soft tissues are currently modeled as isotropic ground matrices with embedded oriented collagen fibers. These tissues behave isotropically in compression as the collagen fibers do not carry compressive load, but anisotropically in tension as the collagen fibers carry load. Current nonlinear bioelastomers like POC, PGS, and PGD have isotropic nonlinear elastic properties in both tension and compression, and typically exhibit strain softening behavior in tension, not the strain stiffening behavior exhibited by soft tissues. Furthermore, introduction of porosity reduces the biomaterial nonlinear response as indicated by the Voight bound (eq. 15) as well as the finite element and experimental results presented in this chapter. If current nonlinear elastic constitutive models are taken as design targets, a significant amount of research is needed to design and fabricate biomaterial scaffolds that can mimic soft tissue nonlinear elastic behavior. In addition, other physical parameters that affect tissue regeneration like permeability and diffusivity become deformation dependent under large strains. Such factors must also be considered when designing scaffolds for soft tissue engineering. In general, little research has been conducted on how to engineer biomaterial scaffolds with nonlinear elastic and mass transport properties. Ironically, the influence of such properties on tissue regeneration can only be assessed once we have the capability to engineer biomaterial scaffolds with nonlinear mechanical and mass transport properties, and can characterize how these properties compare to native soft tissue nonlinear properties.

Acknowledgments

Research reported in this chapter was supported by funding from the National Institutes of Health NIH R01 DE13608, R01 DE016129, R01 DE13416, and R01 AR 053379.

References

1. Checa S, Prendergast PJ (2010) Effect of cell seeding and mechanical loading on vascularization and tissue formation inside a scaffold: a mechano-biological model using a lattice approach to simulate cell activity. J Biomech 43:961-968.
2. Cheng G, Markenscoff P, Zygourakis K (2009) A 3D hybrid model for tissue growth: the interplay between cell population and mass transport dynamics. Biophys J 97:401-414.
3. Khayyeri H, Checa S, Tagil M, O'Brien FJ,Prendergast PJ (2009) Tissue differentiation in an *in vivo* bioreactor: in silico investigations of scaffold stiffness. J Mater Sci Mater Med
4. Brekke JH, Toth JM (1998) Principles of tissue engineering applied to programmable osteogenesis. J Biomed Mater Res 43:380-398.
5. Hollister SJ (2005) Porous scaffold design for tissue engineering. Nat Mater 4:518-524.
6. Hollister SJ, Liao EE, Moffitt EN, Jeong CG, Kemppainen JM (2009) Defining design targets for tissue engineering scaffolds, in Fundamental of Tissue Engineering, Meyer U, Meyer T, Handschel J, Wiesmann HP eds. Springer.
7. Hutmacher DW (2001) Scaffold design and fabrication technologies for engineering tissues–state of the art and future perspectives. J Biomater Sci Polym Ed 12:107-124.
8. Yaszemski MJ, Payne RG, Hayes WC, Langer R,Mikos AG (1996) Evolution of bone transplantation, molecular, cellular and tissue strategies to engineer human bone. Biomaterials 17:175-185.
9. Fung YC (1967) Elasticity of soft tissues in simple elongation. Am J Physiol 213: 1532-1544.
10. Fung YC, Fronek K, Patitucci P (1979) Pseudoelasticity of arteries and the choice of its mathematical expression. Am J Physiol 237:H620-31.
11. Fung YC (1981) The lung–a perspective of biomechanics development. J Biomech Eng 103:91-96.
12. Humphrey JD (2002) Cardiovascular solid mechanics: cells, tissues and organs. Springer-Verlag. New York.
13. Holzapfel GA (2000) Nonlinear solid mechanics: a continuum approach for engineering, John Wiley & Sons, England
14. Ogden RW (1972) Large deformation isotropic elasticity: on the correlation of theroy and experiment for incompressible rubberlike solids. Proc. Royal Soc. London. A326:565-584.
15. Holzapfel GA, Gasser TC, Ogden RW (2000) A new constitutive framework for the arterial wall mechanics and a comparative study of material models. J. Elasticity, 61:1-48.
16. Holzapfel GA, Gasser TC, Ogden RW (2004) Comparison of a multi-layer structural model for arterial walls with a Fung-type model and issues of material stability. J. Biomech. Eng, 126:264-275.
17. Holzapfel GA, Ogden RW (2009) Constitutive modelling of passive myocardium: a structurally based framework for material characterization, Phil Trans. Royal Soc. 367:3445-3475.
18. Humphrey JD, Strumpf RK, Yin FCP (1990) Determination of a constitutive relation for passive myocardium: I. a new functional form. J. Biomech. Eng, 112:333-339.
19. Humphrey JD, Strumpf RK, Yin FCP (1990) Determination of a constitutive relation for passive myocardium: II. parameter estimation. J. Biomech. Eng, 112:340-346.
20. Lipton R. (1994) Optimal bounds on effective elastic tensors for orthotropic composites. Proc. Royal Soc. London A. 444:399-410.
21. Ogden RW (1978) Extremum principles in nonlinear elasticity and their application to composites. 1. Theory. Int. J. Solids Struct. 4:265-282.
22. Hashin Z (1985) Large isotropic elastic deformation of composites and porous media. Int. J. Solids Struct. 21:711-720.
23. Lopez-Pamies O, Idart MI (2009) An exact result for the macroscopic response of porous neo-hookean solids. J. Elasticity 95:99-105.

24. Yang J, Webb AR, Ameer GA (2004) Novel citric acid-based biodegradable elastomers for tissue engineering. Adv. Mat. 16:511-516.
25. Wang Y, Ameer GA, Sheppard BJ, Langer R. (2002) A tough biodegradable elastomer. Nat. Biotechnol. 20:602-606.
26. Jeong CG, Hollister SJ (2010) Mechanical, permeability, and degradation properties of 3D designed poly1,8 octanediol-co-citrate. scaffolds for soft tissue engineering. J Biomed Mater Res B Appl Biomater 93B:141-149.
27. Jeong CG, Hollister SJ (2010) A comparison of the influence of material on *in vitro* cartilage tissue engineering with PCL, PGS, and POC 3D scaffold architecture seeded with chondrocytes. Biomaterials 31:4304-4312.
28. Kemppainen JM, Hollister SJ (2010) Tailoring the mechanical properties of 3D-designed polyglycerol sebacate. scaffolds for cartilage applications. J Biomed Mater Res A, 94:9-18.
29. Migneco F, Huang YC, Birla R, Hollister SJ (2009) Poly(glycerol-dodecanoate), a biodegradable polyester for medical devices and tissue engineering scaffolds. Biomaterials. 30:4063-4069.

Computational techniques for selection of biomaterial scaffolds for tissue engineering

S. Checa, C. Sandino, D.P. Byrne, D.J. Kelly, D. Lacroix and P.J. Prendergast

Abstract: Computational tools are nowadays an indispensable tool for engineering design. Tissue engineering is an interdisciplinary field which so far has been mainly limited to experimental investigations. Here we present several examples of how computer models can be used in the design of tissue engineering scaffolds. Investigations of the effect of scaffold porosity, dissolution rate and/or scaffold material properties on processes such as cell differentiation, migration and angiogenesis are presented. Current limitations and future perspectives in the development and potential application of these models are described.

1 Introduction

Tissue engineering is an interdisciplinary field which holds great potential to revolutionize medicine. It may be defined as a process that aims to regenerate, maintain or improve tissue function by combining cells, a 3D matrix (scaffold) and biochemical factors [1]. According to Vacanti [3], first steps in the field date back to the early 1970s, when Dr. W.T. Green undertook a number of experiments to generate new cartilage using chondrocytes seeded onto spicules of bone and implanted in nude mice [2]. Although his experiments were unsuccessful, he concluded that with the advent of biocompatible materials it would be possible to generate new tissue by seeding viable cells onto appropriately configured scaffolds. Since then, extensive research has been undertaken on the design of scaffolds to establish which scaffold provides the best environment for the cells to proliferate, differentiate and synthesize the new extracellular matrix. Numerous scaffold properties can be manipulated, including the type of material utilised, the shape and size of pores in which cells are located, the mechanical integrity of the

S Checa, DP Byrne, DJ Kelly, PJ Prendergast
Trinity Centre for Bioengineering, School of Engineering, Trinity College, Dublin, Ireland
email: sara.checa@gmail.com, dbyrne24@tcd.ie, kellyd9@tcd.ie, pprender@tcd.ie

C Sandino and D Lacroix
Institute for Bioengineering of Catalonia, Barcelona, Spain
email: clara.sandino@upc.edu, dlacroix@ibec.pcb.ub.es

P.R. Fernandes and P.J. Bártolo (eds.), *Advances on Modeling in Tissue Engineering,* Computational Methods in Applied Sciences 20, DOI 10.1007/978-94-007-1254-6_4, © Springer Science+Business Media B.V. 2011 construct, substrate coating aimed towards promotion of cell adhesion and the incorporation of chemicals and growth factors [4, 5]. In addition, following selection of a scaffold with desirable characteristics, cells can be seeded and conditioned, *in vitro*, through the selection of an appropriate bioreactor system [6]. As in other fields of engineering design, computational tools may play an essential role on the design of scaffolds; the large number of variables involved in the design of tissue constructs could be investigated and even optimized using computational models. Computer simulations of the cellular processes taking place inside porous scaffolds could lead to a fundamental understanding of how these processes are regulated by scaffold properties, such as porosity or dissolution rate, or the surrounding mechanical environment.

Pauwels [7] proposed the first ideas on how the tissue regeneration process is related to mechanical factors (see Prendergast et al. [8] for a review). He observed that new bone formation during fracture repair only occurred under well stabilized mechanical conditions, which are usually provided by an initial formation of soft connective tissues. These ideas were later quantified by Claes and Heigele [9] with the use of computer models. Using finite element models to quantify the distribution of biophysical stimuli in a fracture gap, they used histological sections to quantify a relationship between the magnitude of these stimuli and the distribution of the tissues. Taking into account the biphasic composition of soft tissues, as constituted by a fluid phase embedded in a solid matrix, Prendergast et al. [10] elaborated a theory where stem cell differentiation is regulated by the fluid flow over the cell and the shear strain acting on the cells.

These theories have been implemented in computer models to investigate tissue formation and regeneration in situations such as fracture healing [11, 12, 13, 14, 15, 16], bone/implant interfaces [17, 18, 19, 20, 21, 22], distraction osteogenesis [23, 24, 25] and *in vivo* bone chambers [26, 27, 28]. However, little research has been reported on computational design of scaffolds from the mechanobiological standpoint. Mechanobiological models consist, in general, of two components; the mechanical and the biological/cellular component. The mechanical component of the model aims to determine the mechanical stimuli at the injured site or in a bioreactor, in the case of *in vitro* experiments, and how these stimuli are transmitted to the cells through the scaffold. Finite element analysis is generally used to determine these local mechanical stimuli, being micro-Finite Element models a powerful tool for investigations of scaffolds of irregular geometry. Due to its complexity, the biological part of the computational model has attracted much attention, and two different approaches have evolved for the simulation of the different cellular events taking place during the regeneration of the tissue; a continuum approach and a discrete (lattice modelling) approach. In this chapter, we review these two approaches in their application to the field of tissue engineering together with the use of micro-Finite Element techniques to determine local mechanical stimuli within scaffolds of complex geometries. Finally, we discuss the current challenges of mechanobiological models for tissue regeneration and the potential benefits they may bring to the field of tissue engineering design.

2 Micro-Finite Element models for the determination of mechanical stimuli within tissue engineering constructs

2.1 Overview

Cells seeded in a scaffold attach to the material walls. Their mechanical stimulation depends on the mechanical load applied to the scaffold, the scaffold morphology and the specific location of the cell within the construct. Therefore, the modeling of the scaffold real shape, considering the difference between the solid and the pore phases, is critical to compute the local mechanical stimulus at each specific point, i.e. node of the structure, and to determine the mechanical stimulus transmitted to each cell. For these type of studies, micro-Finite Element models are necessary to account for the real surface of the scaffold.

Two types of mechanical loading of the scaffold are often considered: perfusion fluid flow and compressive loading. Perfusion fluid flows are used in *in vitro* experiments to transport cells, oxygen and nutrients through the scaffold and to stimulate cells through the development of fluid shear stress. Compressive loads can be applied to the scaffolds *in vitro* or *in vivo* to transmit mechanical stimuli to the cells in two different ways: first, through the deformation of the scaffold walls; and second, through the fluid flow within the interconnected pores driven by the deformation of the scaffold.

2.2 Micro-Computed tomography based models.

Most of the scaffolds used in tissue engineering have an irregular geometry with a sample specific porosity. The scaffold morphology will vary depending on the material and on the fabrication method. For example, the macro porosity in a Calcium Phosphate (CaP) based cement can be achieved with the use of a foaming agent. Other scaffolds based on polymers are made by the solving casting particle leaching method. These scaffolds have irregular pore size, pore interconnectivity and wall thickness. The real 3D geometry of the scaffolds can be reconstructed in a non-destructive manner using micro-computed tomography images.

Following this method, the effect of a compressive load in a macroporous CaP bone cement and a porous CaP glass (30% of porosity) was studied by Lacroix et al. [29]. Similar strain distributions under strain controlled loading were predicted for both materials; however, under load controlled loading the stress distribution varied between them. Parts of the scaffold underwent higher strains than others, suggesting that cell differentiation should be different from one part to another. The analysis of perfusion fluid flow within the interconnected pores of these scaffolds was performed by Sandino et al. [30]. Abrupt changes of fluid velocity, pressure and shear stress within the scaffold were predicted due to the low pore interconnectivity of these materials (20% of pore interconnectivity). The fluid velocity and hence the fluid shear stress was almost nil in some pores while peaks of fluid velocity of 1,000 times the magnitude of the inlet fluid velocity were predicted between low interconnected pores inside the scaffold.

A polylactic acid (PLA)-glass composite scaffold with a high porosity (95% of porosity) was studied by Milan et al. [31] under two different mechanical conditions, a perfusion fluid flow and a dynamic compression. In the perfusion fluid flow simulation, the fluid achieved all the interconnected pores of the scaffold. The velocity increased 7.8 times the magnitude of the inlet fluid velocity in the center of the scaffold. The pressure drop between the inlet and the outlet side of the flow was uniform. In the compressive strain simulations, a hetero-geneous strain distribution was predicted (Fig. 1). Most of the scaffold walls were only displaced without being significantly deformed, while strains of 400 times the magnitude of the total strain applied were predicted in some walls suggesting local material damage.

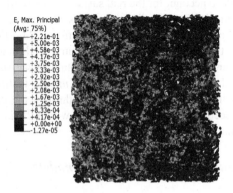

Fig. 1. Maximal principal strain distribution within the PLA-glass scaffold when 5% of compressive strain was applied. A higher magnitude of strain was predicted in the periphery than in the center of the scaffold. Most of scaffold walls were not deformed, since high magnitudes of strain were predicted in some regions.

The comparison of the results between these two scaffolds of different porosity and interconnectivity clearly shows the effect of the morphology on the mechanical stimuli effectively transmitted to the cells inside the scaffold. The critical stimulus for the CaP scaffolds was the fluid flow, presenting nil fluid velocity in some regions and peaks of fluid velocity in others. These peaks of fluid flow can detach the cells from the scaffold surface undergoing apoptosis. The critical stimulus for the PLA-glass scaffolds was the compressive load, presenting stimuli almost nil in some walls and peaks of stimuli in others. These peaks of stress/strain can break the scaffold walls. These results showed how increasing the porosity and interconnectivity of the scaffold, increased the accessibility of the fluid flow and hence the nutrients and oxygen available to the cells, however the mechanical stability of the construct under compressive loads decreased. These results suggest that in cases where perfusion fluid is necessary to seed and stimulate cells, the use of PLA scaffolds can be more effective than the use of CaP scaffolds, while in cases where the scaffold has to support mechanical compressive loads, better results can be obtained using CaP scaffolds.

2.3 Benefits of micro-Finite Element models

Using Micro-Finite Element models, the specific mechanical stimuli transmitted to the cells can be predicted. In the above presented cases of micro-Computed tomography based models, the mechanical stimuli were not homogeneous and were much higher inside the scaffold than the mechanical stimuli imposed to the construct. These results, opposed to what would have been obtained using a continuum homogeneous porosity, demonstrate the importance of modeling the real architecture of the scaffold.

Micro-Finite Element models contribute to the understanding of the relation between the mechanical stimuli distribution and the scaffold architecture. Combined with *in vitro* and *in vivo* experiments, they can be useful in the design of scaffolds. Additionally, according to the mechanical loads applied to the scaffold and the magnitude of mechanical stimuli required within the construct, recommendations about adequate scaffold types can be made.

3 Continuum modelling of cellular processes

3.1 Overview

Understanding the different cellular events that take place within the scaffolds and how they can be influenced by the surrounding environment, e.g. mechanical loading, is essential for the development of tissue engineering strategies. Here computational models may play an important role. Several methodologies have been developed for the simulation of the different cellular events within the tissue engineered constructs. The continuum approach uses differential equations to describe the biological processes that take place during the regeneration of the tissue, such as the migration, the proliferation and the differentiation of cells. In this approach, cells are monitored in terms of a cell density variable that changes in space over time. Diffusion equations are used to simulate the movement of the cells at the injured site. Cell movement is then dictated by the following equation:

$$\frac{\mathrm{d}n}{\mathrm{d}t} = D\nabla^2 n$$

were n is the cell density and D the diffusion coefficient. Cell proliferation and apoptosis are then implemented using specific mathematical functions that try to describe the dynamics of the processes and their dependency on different environmental factors. For example, Kelly and Prendergast [32] described the population of the cells at the injured site to be a combination of migration, proliferation and apoptosis that followed the following equation:

$$\frac{dn}{dt} = D\nabla^2 n + P(S)n - K(S)n$$

where n denotes the number of cells, D is the diffusion coefficient, $P(S)$ is a proliferation rate and $K(S)$ is a cell death rate (either necrosis or apoptosis), as a function of the mechanical stimulus S. In this case, a quadratic relationship was assumed between proliferation/apoptosis and the mechanical stimulus (octahedral shear strain) that can be expressed as:

$$P(S)n - K(S)n = a + bS_0 + cS_0^2$$

Separate equations can theoretically be introduced to describe the specific behavior of the different cell types [32]; for example diffusion coefficients can be implemented as being cell-phenotype specific. In addition, other equations can be used. For example, Bailon-Plaza and van der Meulen [33] and Geris et al. [13] implemented logistic growth functions whereby the proliferation of the cells was dependent on the surrounding matrix density. Gomez-Benito et al. [34] considered the proliferation of the cells to be proportional to the mechanical stimulus, taken as the second invariant of the deviatoric strain tensor.

3.2 Application to tissue engineering

Little has been done on the development of mechanobiological models for tissue regeneration applied to the design of scaffolds in tissue engineering. Early works correspond to Kelly and Prendergast [32] who investigated the effect of mechanical loading on osteochondral defect healing and the benefits of implanting a scaffold at the defect site [35].

Using a continuum approach to simulate cellular processes, they showed an initial shielding of the load transmitted to the defect area which led to early osteogenesis at the injured site. A later load carrier capacity of the newly formed bone resulted in significant amounts of cartilage and fibrous tissue at the injured site [32]. They also demonstrated that the implantation of a scaffold at the defect site can significantly improve the healing of the defect (Fig 2).

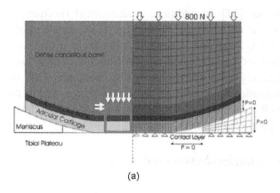

(a)

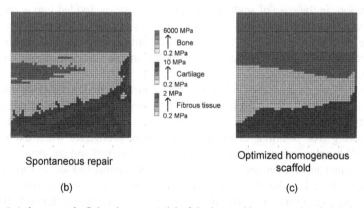

Spontaneous repair

Optimized homogeneous
scaffold

(b)   (c)

Fig. 2. a) Axi-symmetric finite element model of the knee with an osteochondral defect. Right: finite element mesh illustrating loading and boundary conditions. Left: defect (red box) showing origin of mesenchymal stem cells (yellow arrows). b) Predicted patterns of tissue differentiation in a 7 mm osteochondraol defect with no scaffold. Simulations predicted significant amounts of fibrous tissue in the articular cartilage region c) Repair in the presence of a homogeneous scaffold. The tissue construct resulted in a better healing of the defect with a homogeneous articular area almost completely filled with cartilage (Adapted from Kelly and Prendergast, 2006, Ref. 35)

4 Lattice-based models to simulate cellular activity

The continuum approach uses differential equations to simulate the different cellular events. However, diffusion equations assume that cells attempt to achieve an homogeneous distribution within the regenerating region while, in reality, cells migrate following a random walk like behaviour [36] and they may proliferate in a non-uniform manner [37]. In addition, mathematical functions, e.g. a quadratic polynomial, involve of a set of parameters with no physical meaning since parameters are set to best match a set of experimental data. The lattice model approach was adopted in mechanobiological models in order to overcome these limitations. Originally proposed by John von Neumann [38] in the 1950's as models of self-reproducing organisms, nowadays lattice models are applied in the simulation of biological systems, physical phenomena (heat-flow and turbulence) or the design of parallel computers.

4.1 Overview of lattice-based modelling

The idea behind the lattice-based approach is to be able to incorporate the cellular events that individually each cell carries out over time, using a simple set of rules describing the dynamics of the cell. To accomplish this, the area of tissue regeneration, in the case of scaffolds the porous space, is divided into a very fine grid (named a lattice), in which each of the positions (lattice points) represents a possible position for a cell to occupy. In this lattice, cells migrate, proliferate, differentiate and synthesize new extracellular matrix as a response to their surrounding environment.

Here, cell migration and proliferation can be implemented using a random walk approach [21]. Basically, cells migrate through random "jumps" to neighbouring free lattice points and proliferate through mitosis where new cells occupy the neighbouring free positions at random. Cells only proliferate when exposed to favourable biophysical stimuli for the phenotype of the cell and undergo apoptosis when the mechanical stimulus surrounding the cell is either too low or too high for the specific cell; i.e. chondrocytes proliferate under mechanical stimulus favourable for cartilage formation while undergo apoptosis when the mechanical stimulus surrounding them changes to bone formation levels.

Each cell then differentiates based on a set of rules, such as mechanoregulatory rules [10] and/or rules related to the availability of oxygen and nutrients [18]. Cell differentiation can be implemented in a random manner, i.e. not all the cells differentiate at the same time but only a fraction of them do so, and they are chosen randomly. The same principle applies for the proliferation and apoptosis of the cells.

The synthesis of new extracellular matrix by the different cell phenotypes is implemented as a change in the mechanical properties of the finite elements. Since different cell phenotypes can occupy the same element, a rule of mixtures is implemented to determine the material properties of each individual element of the finite element model.

In addition, the lattice model approach allows for the simulation of processes such as angiogenesis, which plays a key role on the regeneration of bone. Blood vessels supply cells with oxygen and nutrients which are essential for their survival and proliferation. Since the diffusion of oxygen is limited to few hundred micrometers from the vessels, the morphology of the new vascular network plays an essential role on the differentiation of the tissues. Capillaries can be simulated as a sequence of endothelial cells which grow and branch following a set of rules [18]. For example, growth factor stimulation of the angiogenic process can be implemented as a tendency for the vessels to grow towards regions of high concentration of vascular endothelial growth factors (Fig. 3).

Fig. 3. A capillary forming from a parent vessel and growing towards a region with high concentration of growth factors (green cube). Rules are implemented for growth, branching and directional movement to simulate the process of angiogenesis.

4.2 Application to tissue engineering

The lattice-based concept was first applied in tissue engineering by Byrne et al. [39] They investigated the effect of scaffold porosity, degradation rate and scaffold mechanical properties on the tissue formation process inside a regular structured bone scaffold (Fig. 4)

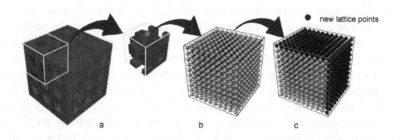

Fig. 4. (a) FE model of a 50% porous scaffold with regular porosity (green). Only one-eighth needs to be modelled because of symmetry (yellow box). The cavity is initially occupied by granulation tissue (in red). (b) Lattice generated for each granulation element to model cellular activity, (c) increase lattice points to account for dissolution of scaffold material (adapted from Byrne et al. 2007, Ref. 39).

In this study, they were able to identify optimal scaffold properties that would lead to the highest amount of bone formation. They showed that the rate of dissolution can either have a positive or negative effect on the amount of bone formation depending on the initial porosity and mechanical strength of the biomaterial. Dissolution of the scaffold increases the porous volume available for bone formation; however, it can also compromise the structural integrity of the scaffold due to a reduction in stiffness and strength (Fig. 5).

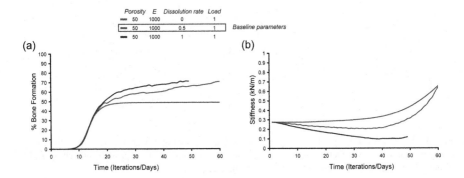

Fig. 5. Effect of dissolution rate on (a) the percentage bone formation and (b) the mechanical stiffness of the bone-scaffold system (Adapted from Byrne et al. 2007, Ref. 39). In the case of the scaffold with 50% porosity, E=1 GPa, and a dissolution rate of 1% per iteration the simulation does not complete because the scaffold collapses under the load because the dissolution rate is too rapid.

These investigations assumed a well vascularised scaffold, with no limitations of oxygen and nutrient supply to the cells. However, vascularisation of large scaffolds remains one of the main challenges in tissue engineering, since the invasion of the tissue construct by blood vessels from the host is often too slow to provide adequate nutrients and oxygen to the cells in the center of the scaffold; resulting in peripheral tissue formation. Investigations on the angiogenic process inside the previously described scaffold were carried out by Checa and Prendergast [18]. Using the lattice-based approach, they simulated the formation of new capillaries inside the tissue construct and its effect on the formation of the tissues. They showed that the initial cell seeding conditions had a significant effect on the vascularisation of the scaffold. A high number of cells initially seeded in the scaffold, resulted in limited penetration of the vascular network and peripheral formation of bone (Fig. 6).

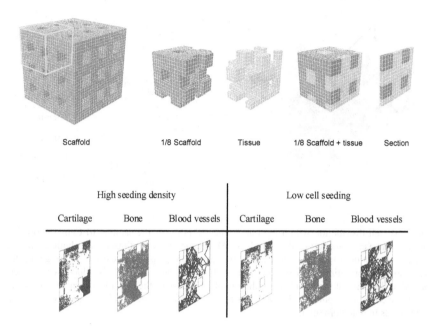

Fig. 6. Cell distribution in a cross section through the scaffold after 60 days. A high seeding density resulted in limited vascularisation and peripheral formation of bone (adapted from Prendergast et al. 2009, Ref. 8).

The above investigations were on printed-type scaffolds; characterized for a regular and structured geometry which leads to uniform distributions of the mechanical stimuli within the pores. However, scaffolds of irregular architecture transmit mechanical loading in a non-homogeneous way, being able to present within a single pore very low and high levels of mechanical stimulation (see section 2). Using micro-CT images to reconstruct scaffold geometry, simulations of tissue differentiation within irregular scaffolds are possible, although they can be computationally very expensive. Simulations in a cylindrical section (1 mm diameter, 0.6 mm height) of a glass based Calcium Phosphate scaffold show that due to the irregular architecture of the scaffold, cell distribution patterns are non-homogeneously distributed within the pores of the scaffold (Fig. 7). In addition, due to a poor interconnectivity of the sample and a small pore size, vascularization is limited to the edges of the construct; which emphasizes the importance of scaffold architecture when attempting to guide the cellular response.

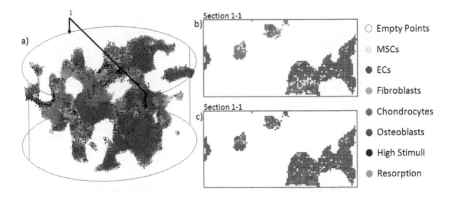

Fig. 7. a) Cell distribution within the interconnected pores of the sample. b) Cells distribution within a transversal section of the sample. c) Mechanoregulatory stimulus distribution within a transversal section of the sample. Since the morphology of the scaffold is irregular, the mechanoregulatory stimulus S was not uniformly distributed. Most of the blood vessels were formed in the pores at the periphery of the sample. Because of the lack of oxygen, chondrocytes are predicted in some pores at the center of the scaffold where the mechanoregulatory stimulus corresponds to bone formation.

4.3 Benefits of lattice approach

The above investigations demonstrate the potential of lattice-based mechano-biological models for their application on the design of scaffolds in tissue engineering. In such studies, as previously discussed, parametric studies can be performed to investigate the effect of scaffold design parameters, such as porosity, stiffness and dissolution rate, on tissue growth and its interaction with the surrounding mechanical environment.

The lattice-based approach uses simple rules based on experimental observations to simulate the behaviour of the cells. Parameters involved using this approach represent physical observations that can, in principle, be experimentally determined, such as the proliferation rate of the cells or the migration velocity. Cellular processes do not occur in a uniform manner, i.e. not all the cells proliferate at the same time or migrate in the same direction. These observations can be easily incorporated using the lattice model approach presented here. Furthermore, complex processes such as angiogenesis can be incorporated. All these cellular events are in addition regulated by multiple paracrine and autocrine signals, which are at the moment the focus of many investigations and could be incorporated in future developments of these models. Finally, as Fig. 7 shows, the lattice modelling approach is readily applicable to intricate and complex scaffold geometries.

5 Discussion and conclusions

Although a robust and promising tool, common to all these models is the issue of validation, essential before they can be used as a decision-support tool for in tissue engineering and regenerative medicine [40]. In general, they have all shown good qualitative agreement with experimental observations; however their predictive capabilities may be questioned due to the large number of parameters they involve. A model is as good as its input values and, even today, there is still not a great amount of experimental data to feed these models [41]. For example, it is widely accepted that cell migration or proliferation is affected by the mechanical environment; however how this regulation happens remains a subject of intense experimental investigation. To us one of the advantages of the lattice modeling approach is that the simulation can be considered to consist of a set of algorithms describing the various mechanoregulated processes (see Fig. 6), and that each of these mechanoregulated processes can be cast in terms of algorithms that may be validable individually.

Another aspect that arises when trying to corroborate these models using *in vivo* experimental observations is the fact that *in vivo* experiments often show a great variability between specimens. The simulation method presented above is stochastic because it incorporated random cell activities (based on Monte Carlo modeling in this work). Khayyeri et al. [28] investigated whether this stochasticity could explain inter-specimen variability in the tissue differentiation process as seen experimentally in an *in vivo* bone chamber, either on its own or when variability in environmental factors is included. They concluded that, although these factors had an influence on the simulations outcome, they could not explain the variability between the animals to the same extent as seen in the experiments and suggested that genetic factors may play a key role in the variability observed in animal experiments and, therefore, between patients in the clinic as well. Investigations on why and what factors contribute to this variability are necessary before they can be implemented in computer models.

In conclusion, considerable advances are being made in the development of computational tools for the design of scaffolds for tissue engineering applications. These techniques allow for analysis of the individual cellular processes taking place within the construct as well as of the interactions within these processes. Simulations of tissue formation inside porous constructs hold a great potential for the development and design of new strategies in tissue engineering which combined with experimental investigations could lead to significant advances in our abilities to perform engineering design in tissue engineering applications.

Acknowledgments

The research reported here is funded by a Science Foundation Ireland Principal Investigator grant to P.J. Prendergast

REFERENCES

1. Langer R, Vacanti JP (1993) Tissue engineering. Science; 260:920-6.
2. Green WT Jr. (1977). Articular cartilage repair. Behavior of rabbit chondrocytes during tissue culture and subsequent allografting. Clin Orthop Relat Res; 124:237-50.
3. Vacanti JP (2006). History of Tissue Engineering and a glimpse into its future. Tissue Engineering, 12:1137-1142.
4. Howard D, Buttery LD, Shakesheff KM, Roberts SJ (2008) Tissue engineering: strategies, stem cells and scaffolds. Journal of Anatomy, 213:66-72.
5. Hutmacher DW, Schantz JT, Lamm CX, Tanm KC, Lim TC (2007) State of the art and future directions of scaffold-based bone engineering from a biomaterials perspective. J Tissue Eng Regen Med., 1(4):245-60.
6. Martin I, Wendt D, Heberer M (2004) The role of bioreactors in tissue engineering. Trends Biotechnol, 22:80-86.
7. Pauwels F (1973) Atlas zur Biomechanik der gesunden und kranken Hüfte. Ed. 1 Berlin, Heidelberg, New York, Springer Verlag.
8. Prendergast PJ, Checa S, Lacroix D (2009) Computational models for tissue differentiation. In: Computational Methods in Biomechanics, (Editors: Suvranu De, Farsh Guilak, Mohammad Mofrad), Springer: New York, In Press.
9. Claes LE, Heigele CA (1999) Magnitudes of local stress and strain along bony surfaces predict the course and type of fracture healing. Journal of Biomechanics 32:255-266.
10. Prendergast PJ, Huiskes R, Søballe K (1997) ESB Research Award 1996. Biophysical stimuli on cells during tissue differentiation at implant interfaces. Journal of Biomechanics 30:539-548.
11. Loboa EG, Beaupré GS, Carter DR, (2001). Mechanobiology of initial pseudarthrosis formation with oblique fractures. Journal of Orthopaedic Research, 19(6):1067-1072.
12. Andreykiev A, van Keulen F, Prendergast PJ (2008) Simulation of fracture healing incorporating mechanoregulation of tissue differentiation and dispersal/proliferation of cells, Biomechanics and Modeling in Mechanobiology 7: 443-461.
13. Geris L, Gerisch A, Vander Sloten J, Weiner R, Oosterwyck HV (2008) Angiogenesis in bone fracture healing: a bioregulatory model. J. Theor. Biol. 7:137-158.
14. Hayward LNM, Morgan EF (2009) Assessment of a mechano-regulation theory of skeletal tissue differentiation in an *in vivo* model of mechanically induced cartilage formation. Biomechanics and Modeling in Mechanobiology. DOI 10.1007/s10237-009-0148-3.
15. Isaksson H, Wilson W, van Donkelaar CC, Huiskes R, Ito K (2006) Comparison of biophysical stimuli for mechano-regulation of tissue differentiation during fracture healing. J. Biomech. 39:1507-1516.
16. Lacroix D, Prendergast PJ (2002) Three-dimensional simulation of fracture repair in the human tibia, Computer Methods in Biomechanics and Biomedical Engineering 5:369-376
17. Andreykiv A, van Keulen F, Prendergast PJ (2008) Computational mechanobiology to study the effect of surface geometry on peri-implant tissue differentiation. Journal of Biomechanical Engineering 130, Paper No 051015.
18. Checa S, Prendergast PJ (2009) A mechanobiological model for tissue differentiation that includes angiogenesis: a lattice-based modeling approach. Ann. Biomed. Eng. 37:129-145.
19. Huiskes R, van Driel WD, Prendergast PJ, Soballe K (1997) A biomechanical regulatory model of peri-prosthetic tissue differentiation. J. Mater. Sci. Mater. Med. 8:785-788.
20. Liu X, Niebur G (2008). Bone ingrowth into a porous coated implant predicted by a mechano-regulatory tissue differentiation algorithm. Biomechanics and Modeling in Mechanobiology, 7(4):335-344.
21. Perez MA, Prendergast PJ (2007) Random-walk models of cell dispersal included in mechanobiological simulations of tissue differentiation. J. Biomech. 40:2244-225.
22. Prendergast P, Huiskes R, (1996) Finite element analysis of fibrous tissue morphogenesis: a study of the 'osteogenic index' using a biphasic approach, Mechanics of Composite Materials 32:209-218.

23. Boccaccio, A., Pappalettere, C., Kelly, D. J. (2007) The influence of expansion rates on mandibular distraction osteogenesis: a computational analysis, Annals of Biomedical Engineering 35:1940-60.

24. Boccaccio A, Prendergast PJ, Pappalettere C, Kelly DJ (2008) Tissue differentiation and bone regeneration in an osteotomized mandible: a computational analysis of the latency period, Medical and Biological Engineering and Computing 46:283-298.

25. Isaksson H, Comas O, van Donkelaar CC, Mediavilla J, Wilson W, Huiskes R, Ito K (2007) Bone regeneration during distraction osteogenesis: mechano-regulation by shear strain and fluid velocity. J. Biomech. 40:2002-2011.

26. Geris, L., Vandamme, K., Naert, I., Vander Sloten, J., Duyck, J., Van Oosterwyck H. (2008) Application of mechanoregulatory models to simulate peri-implant tissue formation in an *in vivo* bone chamber. J. Biomech. 41:145-154.

27. Geris L, Andreykiv A, Van Oosterwyck H, Vander Sloten J, van Keulen F, Duyck J, Naert I (2004) Numerical simulation of tissue differentiation around loaded titanium implants in a bone chamber. Journal of Biomechanics 37:763-769.

28. Khayyeri H, Checa S, Tagil M, Prendergast PJ (2009) Corroboration study of mechano-biological simulations of tissue formation in an *in vivo* bone chamber using a lattice-modeling approach. Journal of Orthopaedic Research. DOI 10.1002/jor.20926.

29. Lacroix D, Château A, Ginebra MP and Planell JA (2006) Micro-finite element models of bone tissue-engineering scaffolds. Biomaterials 27:5326-5334.

30. Sandino C, Planell JA, Lacroix D (2008) A finite element study of mechanical stimuli in scaffolds for bone tissue engineering. Journal of Biomechanics 41:1005-1014.

31. Milan JL, Planell JA, Lacroix D (2009). Computational modelling of the mechanical environment of osteogenesis within a polylactic acid–calcium phosphate glass scaffold Biomaterials 30:4219-4226.

32. Kelly DJ, Prendergast PJ (2005) Mechano-regulation of stem cell differentiation and tissue regeneration in osteochondral defects, Journal of Biomechanics 38:1413-1422.

33. Bailon-Plaza A, van der Meulen MC (2001) A mathematical framework to study the effects of growth factor influences on fracture healing. J. Theor. Biol. 21:191-209.

34. Gomez-Benito MJ, Garcıa-Aznar JM, Kuiper JH, Doblare M (2005) Influence of fracture gap size on the pattern of long bone healing: a computational study. Journal of Theoretical Biology 235 (1):105-119.

35. Kelly DJ, Prendergast PJ (2006) Prediction of the optimal mechanical properties for a scaffold used in osteochondral defect repair. Tissue Eng. 12:2509-2519.

36. Carter SB (1965) Principles of cell motility: the direction of cell movement and cancer invasion. Nature 208 (16):1183-1187.

37. Palsson BO, Bhatia SN (2004) Tissue Engineering. Pearson Prentice. Hall Bioengineering.

38. Von Neumann J (1966). Theory of Self-Reproducing Automata. Univ. of Illinois Press.

39. Byrne DP, Lacroix D, Planell JA, Kelly DJ, Prendergast PJ (2007) Simulation of tissue differentiation in a scaffold as a function of porosity, Young's modulus and dissolution rate: application of mechanobiological models in tissue engineering. Biomaterials 28: 5544-5554.

40. Prendergast PJ (2001). An analysis of theories in biomechanics. Engineering Transactions (Rozprawy Inzynierskie) 49:117-133.

41. Isaksson H, van Donkelaar C, Huiskes R, Ito K (2008) A mechano-regulatory bone-healing model incorporating cell-phenotype specific activity. Journal of Theoretical Biology 252: 230-246.

Modelling bone tissue engineering. Towards an understanding of the role of scaffold design parameters

José A. Sanz-Herrera, Manuel Doblaré and José M. García-Aznar

Abstract Tissue engineering emerged in the beginning of 90's as a new paradigm in medicine and life sciences. A number of successful results have been reported although the clinical practice has not been reached so far. One of the problems encountered in this methodology is the coordination among the different biophysical fields involved and the uncertain behaviour of a specific cell carrier, i.e, *scaffolds* in each application. Moreover, the role of the scaffold design in tissue regeneration is poorly understood and new protocols have to be tested over different tissues. In order to advance in the knowledge of the scaffold behaviour towards its functionality and to reduce animal experimentation, computer simulation may serve as a useful platform for scaffold design, once the models are sufficiently validated. In this framework, the potential of numerical simulation, based on a multiscale and multiphysic approach, is highlighted in this work. Furthermore, the role of scaffold microstructural anisotropy in bone tissue regeneration is analyzed using this approach.

José A. Sanz-Herrera
Group of Structural Mechanics and Materials Modeling, Aragón Insitute of Engineering Research (I3A), University of Zaragoza. CIBER-BBN Centro Investigación Biomédica en Red en Bioingeniería, Biomateriales y Nanomedicina and School of Engineering, University of Seville e-mail: jsanz@us.es

Manuel Doblaré
Group of Structural Mechanics and Materials Modeling, Aragón Insitute of Engineering Research (I3A), University of Zaragoza. CIBER-BBN Centro Investigación Biomédica en Red en Bioingeniería, Biomateriales y Nanomedicina e-mail: mdoblare@unizar.es

José M. García-Aznar
Group of Structural Mechanics and Materials Modeling, Aragón Insitute of Engineering Research (I3A), University of Zaragoza. CIBER-BBN Centro Investigación Biomédica en Red en Bioingeniería, Biomateriales y Nanomedicina e-mail: jmgaraz@unizar.es

P.R. Fernandes and P.J. Bártolo (eds.), *Advances on Modeling in Tissue Engineering*,
Computational Methods in Applied Sciences 20, DOI 10.1007/978-94-007-1254-6_5,
© Springer Science+Business Media B.V. 2011

1 Motivation of numerical simulation on tissue engineering

In the last three decades, tremendous progress and encouraging results have been achieved in the field of tissue engineering. Special interest has been paid by material scientists to engineer scaffolds from natural materials (collagen, alginate, hydroxyapatite) or synthetic polymers (polyglycolide, polyactide, polyactide coglycolide) that mimic the extracellular space. The advances in this area have lead, for instance, to the combination of chondrocytes and appropriate polymer templates of complex 3-dimensional architectures like the human ear. In this work, Cao et al. [1] reported the production of neocartilage after transplantation into athymic mice. However, the choice of the biomaterial and its composition was ad hoc and based on the experience of the authors with unpredictable outcome. Nonetheless, mathematical modelling and numerical simulation allow to isolate effects and phenomena of such events and qualitatively predict the behaviour for a certain application. In fact, for example, Kelly and Prendergast [2] simulated osteochondral defects repair using scaffolds based on a mathematical model of cell/tissue differentiation. Results show the evolution of different cell phenotypes along time.

One of the main challenges in tissue engineering is to regenerate vascularized tissues [3, 4, 5]. A successful product of tissue engineering or organ fabrication requires an adequate source of healthy, expandable cells, friendly scaffolds, and the optimal bioreactors [6]. Potential methods to perform this product may include controlled release of angiogenic factors from scaffolds [7], seeding endothelial cells directly into the scaffold, and engineering the vasculature directly into the tissue using various methods such as microfabrication [8]. One possible approach to overcome the problem of vascularization was presented in Warnke et al. [9]. In this work, the authors repaired a mandible lost by the patient due to a tumoral disease. They used as scaffold a titanium cage filled with mineral blocks, infiltrated with BMP-7 and patient's bone marrow. To get vascularization, they implanted the scaffold into the latissimus dorsi muscle, see Fig. 1. *In-vivo* results showed bone remodeling and mineralization inside the mandibular transplant before and after the transplantation. This is an early example of the use of the human body itself as a bioreactor to promote vascularization. Numerically, vascularization may be tested by evaluating oxygen consumption and the ability of cells to migrate and regenerate blood vessels which may be qualitatively modelled through the scaffold permeability. For example, Cioffi et al. [10] evaluated numerically oxygen distributions over scaffolds in perfusion culture systems. Additionally, Sanz-Herrera et al. [11] performed a numerical study to evaluate both the permeability and mechanical properties of a commercial bioceramic scaffold, where the numerical results were corroborated experimentally (see Tables 1 and 2). These numerical strategies may serve to aid to the scaffold design. In fact, Hollister and co-workers have focused on the design of bone scaffolds as an optimization problem to get a microstructure as similar as possible (mechanical properties, porosity, pore size, etc) to the original tissue in the implanted region. For this purpose, the homogenization theory was extensively applied [12, 13, 14].

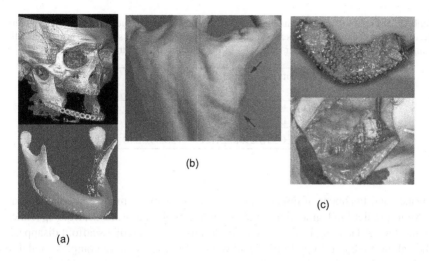

(b)

(c)

(a)

Fig. 1 Mandible repair using the human body as a bioreactor: (a) generation of the defect model in CAD and (b) implantation of the scaffold into the patient's body to serve as a bioreactor. New bone appeared within the scaffold (c) after finally transplanted in the mandible defect. From [9].

Table 1 Intrinsic permeability of the Sponceram® groups (mean value ± standard deviation). Results were obtained both experimentally by means of a permeability test and numerically using homogenization theory over several RVEs of the fluid flow within the scaffold microstructure. From [11].

	Porosity (%)	Experimental $k\ (m^2)$	Numerical $k\ (m^2)$	Bone $k\ (m^2)$
S20-90	90.16±0.95	$3.76 \cdot 10^{-8} \pm 4.38 \cdot 10^{-9}$	$3.49 \cdot 10^{-8} \pm 2.56 \cdot 10^{-9}$	$9.5 \cdot 10^{-9}$
S30-90	88.75±0.36	$3.17 \cdot 10^{-8} \pm 3.62 \cdot 10^{-9}$	$2.31 \cdot 10^{-8} \pm 6.25 \cdot 10^{-10}$	$7.8 \cdot 10^{-9}$
S30-90HA	79.81±2.18	$1.79 \cdot 10^{-8} \pm 4.09 \cdot 10^{-9}$	$1.88 \cdot 10^{-8} \pm 1.25 \cdot 10^{-9}$	$2.1 \cdot 10^{-9}$

Gutierres et al. [15] presented a porous scaffold structure Bonelike® implanted on human 54 years mean age. The application was the treatment of medial compartment osteoarthritis of the knee and it was implanted during osteotomy surgery. The scaffolds were evaluated at 3, 6, 9 and 12 months by scanning electron microscopy and histology. Results revealed vascularization and bone ingrowth within the scaffolds. Recent works [16, 17] have attempted to simulate similar bone tissue engineering applications. A multiscale strategy was used for this purpose concluding that after a concise model validation, the methodology may be very useful to evaluate the availability of a certain scaffold in different applications.

New trends in tissue engineering include the use of nanostructures to promote growth of blood vessels [18]. Also, new challenges comprise understanding fundamental biology as essential for new developments in cell therapy and elucidate the role of signals in cell behavior [5]. Moreover, in parallel to tissue engineering, the use of biomaterials for drug delivery offers many applications for medical

Table 2 Young's modulus of the Sponceram® groups (mean value ± standard deviation). Results were obtained both experimentally in a two-plates compression test and numerically using homogenization theory over several RVEs of the scaffold microstructure. From [11].

	Porosity %	Experimental E (MPa)	Numerical E (MPa)	Bone E (MPa)
S20-90	90.16±0.95	8.73±4.64	8.21±3.06	31.24
S30-90	88.75±0.36	11.03±2.24	9.07±1.43	43.67
S30-90HA	79.81±2.18	29.40±1.21	30.63±2.61	168.43

advance and treatment of diseases [19]. In this scenario, numerical simulation may serve to predict biodegradation kinetics and the evolution of the scaffold specimen along time. Adachi et al. [20] presented a framework to predict scaffold disappearance along time due to hydrolysis pathways. Moreover, a more complex model to evaluate the degradation of biodegradable polymers was presented in [21].

Even though successful results in several applications of tissue engineering have enabled the approve of very few commercial products for clinical application by the US Food and Drug Administration (FDA), companies mostly devoted to tissue engineering have yet to prove themselves economically viable [22, 23]. This is partly because it is still considered an immature and young field [24] with unsuccessful, or even daunting, results in several areas. Nevertheless, given the scientific promise, potential social impact, and the young age of the field, many believe that it should be only a matter of time until tissue engineering reaches the main stream of surgical practice [22]. For this target, numerical simulation may result a useful tool to accomplish this challenge.

In this work, we analyze the effect of scaffold microstructural anisotropy on new bone tissue formation by means of a multiscale numerical model. For this purpose, different scaffold microstructures are analyzed including different degrees of internal anisotropy. The results show that when the scaffold microstructure is properly interconnected and the porosity is high, similar rates of bone regeneration are found. However, scaffold microstructural anisotropy has important consequences in the shape and distribution of the new formed tissue.

2 A multiscale strategy in bone tissue engineering

The need for including a multiscale analysis of bone tissue engineering applications comes from the different biological and physiological processes involved in this methodology. Fig. 2 shows the different scales identified in bone tissue engineering applications: from the left to the right, we can first distinguish the daily–life macroscopic scale where the scaffold is implanted in the bone organ. At this level, the mechanical forces are transferred from the bone organ to the scaffold due to the normal activity of the individual. In this case, new bone tissue regeneration may be

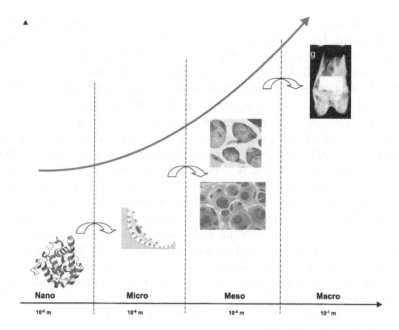

Fig. 2 Different spatio-temporal scales involved in bone tissue engineering applications. Model parameters increase as the model is posed at a more phenomenological (macroscopic) fashion. Some figures taken with permission from [25] and [26].

considered a bone regeneration process, where cell processes, such as, cell migration, vasculogenesis, differentiation and proliferation are involved. All of these phenomena may be modeled following a phenomenological approach. In this case, a number of model parameters are needed such as an effective 'diffusion' parameter to model cell migration and the apparent mechanical properties of the scaffold including its evolution along time as new tissue is growing up in its interior. Going down to the mesoscale (see Fig. 2) the above mentioned macroscopic parameters may be derived by knowing the microstructural geometry of the scaffold. In fact, mechanical stress and strain distributions may also be computed at this meso level. However, phenomena such as cell adhesion or cytoskeletal stress distributions are still out of this scale, being necessary to work at the microscopic level as it is shown in Fig. 2. Finally, a detailed model should include an analysis at the protein level (nanoscale) to try to understand how the biochemical cascade of present signals transduce into mechanical cues and viceversa, i.e, *mechanotransduction*, regulating processes as cell adhesion, proliferation and differentiation. In conclusion, when the model is posed at a lower scale of observation more fundamental phenomena have to be indentified, thus reducing the number of model parameters and consequently the fitting process. Therefore, the modelling of bone tissue engineering problems reduced to one single scale, presents important limitations since interesting phenomena occurs at any scale. However, a multiscale formulation of the problem considering all the scales presented in Fig. 2 is presumably impossible due to computer limitations.

In this work, a two-scales formulation of bone tissue engineering problems is presented. For the sake of simplicity and conventional notation we refer to a micro-macro coupled model. This model here developed for bone tissue regeneration within scaffolds is summarized in Fig. 3 and can also be found in [17, 28]. Briefly, the macroscopic variables are obtained at the macro-scale, $\Omega^0 \equiv \Omega^0_{-S} \cup \Omega^0_S$ by solving the macroscopic equations explained below, where Ω^0_{-S} corresponds to the bone subdomain and Ω^0_S being the scaffold subdomain. Then, the *localization problem* is solved in a representative volume element (RVE) of the microstructure, RVE $\equiv \Omega^\varphi_S \cup \Omega^\varphi_F$, to obtain the microscopic distributions of the variables that regulate the bone growth process. Note the use of superscripts "0" and "φ" for the macro and micro scales, respectively. Moreover, the evolution of both scales depends on time. However, the explicit time dependence will be avoided in the notation for simplicity. Once the microscopic bone growth model and the evolution of the scaffold resorption processes are solved, the mechanical and flow properties are updated (homogenized) taking into account the newly formed bone. Mechanical properties both in the bone organ and scaffold are considered to be fully anisotropic. The loop is closed after passing the effective macroscopic properties needed in the solution of the macroscopic model. Both micro and macro models are explained in detail in the following subsections. Nonetheless, the reader is referred to [17, 28] for additional information on the model.

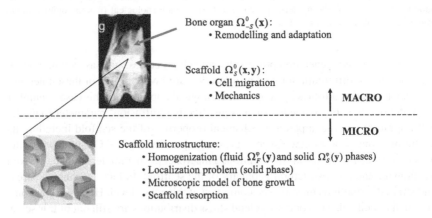

Fig. 3 Two-scale analysis developed for bone ingrowth in scaffolds. The tissue (macro) scale passes information to the pore scale (micro) in the form of macroscopic variables obtained in the macroscopic analysis. At the microscopic scale, the distributions of these macroscopic variables are derived by solving a *localization problem*. Then the bone ingrowth problem is solved and the overall mechanical and flow properties are obtained through *homogenization*. These properties are updated and this information is passed to the macroscopic scale. Figures taken with permission from [25], [26] and [27].

2.1 Numerical analysis of the macroscopic scale

Bone regeneration using scaffolds is a complex process that can be assumed to follow the same sequence of events characteristics of conventional bone regeneration under normal biocompatibility conditions. First, a haematoma is formed and inflammation occurs [29]. At this level the walls of the scaffold are filled with mesenchymal stem cells (MSCs) and preosteoblastic cells. This migration process from the surrounding bone to the scaffold core is here modeled by the Fick's law,

$$\dot{\bar{c}} = \nabla \cdot (\mathbf{D} \cdot \nabla \bar{c}) \quad \text{in} \quad \Omega_S^0(\mathbf{x}, \mathbf{y})$$
$$+\text{boundary and initial conditions} \tag{1}$$

where the diffusion matrix denoted by $\mathbf{D}(\mathbf{x}, \mathbf{y})$ is considered fully anisotropic and $\bar{c}(\mathbf{x}) := c(\mathbf{x})/c_s \in [0, 1]$, being $c(\mathbf{x})$ the number of cells per unit volume in the scaffold domain and c_s the number of cells per unit volume surrounding the scaffold boundary, which is assumed to be the maximum density of MSCs and preosteoblastic cells available, $\bar{c} \leq c_s$. $\mathbf{x}$ denotes the macroscopic scale and $\mathbf{y}$ the microscopic one.

The effect of loading in bone regeneration in Ω_S^0 and bone remodeling in Ω_{-S}^0 is assumed to be dependent on a mechanical stimulus based on the strain energy density (similar to the one used in the classical bone remodeling theory proposed by [30]). Linear elasticity in absence of body forces is assumed at the macroscopic domain,

$$\nabla \cdot \sigma = 0$$
$$\varepsilon = \tfrac{1}{2}(\nabla \mathbf{u} + \nabla \mathbf{u}^T) \quad \text{in} \quad \Omega_{-S}^0(\mathbf{x}) \cup \Omega_S^0(\mathbf{x}, \mathbf{y})$$
$$\sigma = \mathbb{C}^0 \varepsilon$$
$$+\text{boundary conditions} \tag{2}$$

being $\varepsilon(\mathbf{x})$ and $\sigma(\mathbf{x})$ second-order tensor fields denoting the (linearized) strain and stress, respectively, and $\mathbb{C}^0$ the macroscopic fourth-order elasticity tensor defined as,

$$\mathbb{C}^0 = \begin{cases} \mathbb{C}^0(\mathbf{x}) & \text{in} \quad \Omega_{-S}^0(\mathbf{x}) \\ \mathbb{C}^0(\mathbf{x}, \mathbf{y}) & \text{in} \quad \Omega_S^0(\mathbf{x}, \mathbf{y}) \end{cases} \tag{3}$$

On one hand, coupling between micro and macro scales occurs in the scaffold subdomain and is established in terms of the respective diffusion matrices $\mathbf{D}(\mathbf{x}, \mathbf{y})$ and elasticity tensor $\mathbb{C}^0(\mathbf{x}, \mathbf{y})$. The macroscopic diffusion matrix $\mathbf{D}(\mathbf{x}, \mathbf{y})$ is correlated to the Darcy's permeability, which is obtained by means of the homogenization process explained below. In a similar way, the elasticity tensor $\mathbb{C}^0(\mathbf{x}, \mathbf{y})$ in the scaffold subdomain is also evaluated using the homogenization technique. In that approach, the time-evolution of the microstructure due to both bone growth and scaffold resorption in the scaffold subdomain have been taken into account. These two processes occur at the microstructural level and are explained in detail in the following subsections.

On the other hand, the elasticity tensor in the bone subdomain $\mathbb{C}^0(\mathbf{x})$ is macroscopically evaluated by means of an anisotropic bone remodeling computational model [31, 32], able to predict the evolution of bone porosity and anisotropy as a function of the mechanical conditions. In fact, this model is posed by introducing as internal state variables at the macroscopic level: the apparent density and the 'fabric tensor'.

2.2 Microscopic model of bone ingrowth

At the microstructural level, bone growth occurs at the scaffold surface after MSCs attachment and subsequent proliferation, differentiation and bone matrix production. In this context, we assume that bone formation is driven by a mechanical stimulus evaluated at the microstructural level. Since we pose the problem at the microsctructural level, we consider that the activity of bone cells and subsequent bone matrix production and/or resorption happen at the scaffold pore surface. Then, we establish that the rate of bone mass formation at the local scaffold microsurface may be written as,

$$\dot{M} = \dot{M}(\bar{c}, \Psi) \qquad \text{in} \quad \partial\Omega_S^\varphi(\mathbf{y}) \tag{4}$$

where the bone mass rate $\dot{M}$ is a function of a microscopic mechanical stimulus Ψ and the current macroscopic population of MSCs, $\bar{c}$. In fact, we consider that the activity of the cells attached to the scaffold surface starts when a mechanical threshold is achieved. Above this threshold, MSCs start to differentiate into osteoblasts and to deposit bone matrix in the surface while under that value they preferentially stay inactive [33, 34]. Therefore, we can express bone growth on the scaffold surface within the scaffold domain as,

$$\dot{M} = \begin{cases} k_s(\bar{c}) \cdot (\Psi - \Psi^* - w) & \text{if } \Psi - \Psi^* > w \\ 0 & \text{otherwise} \end{cases} \tag{5}$$

where $k_s(\bar{c}) = k \cdot \bar{c}$ is an empirical rate function proportional to the macroscopic concentration of MSCs $\bar{c}$ currently attached to the surface of the scaffold.

The assumed microscopic mechanical stimulus is evaluated at the scaffold pore surface as,

$$\Psi = \left(\sum_{i=1}^{N} n_i \bar{\sigma}_i^m \right)^{1/m}$$

where N is the number of different load cases, n_i the average number of cycles per time unit for the load case i, m a model parameter that quantifies the effect of loading cycles [30] and $\bar{\sigma}$ is a effective scalar stress that according to [30] may be expressed as,

$$\bar{\sigma} = \sqrt{2EW} \qquad (6)$$

being E the local Young Modulus distinguishing among scaffold (s), non-mature bone (n) or mature bone (m). Finally, $W = 1/2\sigma^{\varphi} : \varepsilon^{\varphi}$ is the strain energy density at the scaffold microsurface. This is obtained through the localization problem and subsequent distribution of strains along the scaffold microstructure (the reader is addressed to [17] for further details).

As a first simplified approach, the time evolution of the mechanical properties of non-mature bone is considered to be linear [20],

$$\mathbb{C}_n^{\varphi} = \mathbb{C}_m^{\varphi} \frac{t}{T_m} \quad 0 \leq t \leq T_m \qquad (7)$$

where $\mathbb{C}_i^{\varphi}, i = n, m$ are the fourth order elasticity tensors of non-mature (n) and mature bone (m), respectively. Note that, at the microscopic level, the mechanical properties depend only on the mineralization ratio since the distinction between cancellous and cortical bone has no sense at this level. Newly formed bone matrix is considered at this scale as isotropic as a first approach.

2.3 Microscopic model of the biodegradation kinetics in polymeric scaffolds

For biodegradable scaffolds, the degradation mechanism is modeled as a hydrolysis process. The water content in the polymer chemically reacts and breaks down, and as consequence the biomaterial bulk erosion occurs [35]. We consider that the dimensionless spatial rate of water content d, defined as the ratio between the concentration of water [mol] in the bulk of the polymer and the concentration of water [mol] at the boundary of the scaffold, is governed by the diffusion equation [20],

$$\dot{d} = \alpha \Delta d \quad \text{in} \quad \Omega_S^{\varphi}(\mathbf{y})$$
$$+\text{boundary and initial conditions} \qquad (8)$$

with $\alpha > 0$ the diffusion coefficient. The change rate of the molecular weight of the polymer due to hydrolysis is assumed to depend on the local water content [20] $(0 \leq d \leq 1)$, as

$$\dot{W}(d) = -\beta d \qquad (9)$$

being $\beta > 0$ a material constant. The mechanical properties of the polymer are assumed to be linearly related to its molecular weight [20]

$$\mathbb{C}^{\varphi}(W(t)) = \mathbb{C}^{\varphi}(t = 0)\frac{W(t)}{W_0} \quad \text{in} \quad \Omega_S^{\varphi}(\mathbf{y}) \qquad (10)$$

with W_0 the reference molecular weight of the polymer with associated elasticity tensor $\mathbb{C}^{\varphi}(t = 0)$.

2.4 Variational formulation

Macroscopically, the problem is governed by Eqs. (1) and (2). Nevertheless the constitutive relations of these equations depend on the evolution of the underlying microscopic scale. This evolution is determined by the new bone formation onto the scaffold microstructure (and the closure of pores) and the resorption process of the scaffold. The bone growth model is approached by means of the model presented in section 2.2. On the other hand, the biodegradation of the scaffold is modeled as in section 2.3. Consequently, the weak formulation of the problem read as,

$$\int_{\Omega_S^0} \dot{\bar{c}} \, \eta^0 \, d\Omega_S^0(\mathbf{x}) + \int_{\Omega_S^0} \nabla_x \eta^0 \, \mathbf{D} \, \nabla_x \bar{c} \, d\Omega_S^0(\mathbf{x}) = \int_{\partial \Omega_S^0} \hat{\mathbf{q}} \cdot \mathbf{n} \, \eta^0 \, d\partial \Omega_S^0(\mathbf{x}) \qquad (11)$$

$$\int_{\Omega_{-S}^0} \sigma_0 : \varepsilon_0(\mathbf{w}^0) \, d\Omega_{-S}^0(\mathbf{x}) +$$

$$+ \int_{\Omega_S^0} \varepsilon_0(\mathbf{w}^0) : \left[\frac{1}{|\Omega^\varphi|} \int_{\Omega_S^\varphi} \mathbb{C}^\varphi \varepsilon^\varphi(\mathbf{u}^\varphi) \, d\Omega_S^\varphi(\mathbf{y}) \right] d\Omega_S^0(\mathbf{x}) = \int_{\partial \Omega_{-S}^0} \hat{\mathbf{t}} \cdot \mathbf{w}^0 \, d\partial \Omega_{-S}^0(\mathbf{x})$$

$$\forall \, \mathbf{w}^0(\mathbf{x}), \eta^0 \in V_x \qquad (12)$$

being the space of admissible variations defined as,

$$V_x := \left\{ \mathbf{w}^0(\mathbf{x}), \eta^0(\mathbf{x}) \in \mathbb{R}^3 \mid w_i^0, \eta^0 \in H^1(\Omega_S^0) \text{ and } \mathbf{w}^0 = \mathbf{0}, \eta^0 = 0 \text{ on } \partial \Omega_S^0 \right\} \qquad (13)$$

with $H^1(\Omega^\varepsilon)$ the first-order Sobolev space. In Eq. (11) $\mathbf{D}$ is correlated with the permeability matrix, although $\bar{c}$ and $\mathbf{v}^\varphi$ (fluid velocity) are not constitutively related. Therefore, $\mathbf{D}$ is obtained once solving the localization problem associated to the microscopic fluid phase to obtain the permeability matrix $\mathbf{K}$ and assuming $\mathbf{D} = \mathbf{D} \cdot D_0$, being D_0 a model parameter. See ref. [17] for further details.

   As simplification, we have considered that the bone organ and the scaffold are connected, i.e., there not exists mechanical interface between them. Moreover, the free boundary of the scaffold (not connected with the rest of the bone organ) does not support boundary loads. After a homogenization procedure in (12) we get the two-scales variational form of the mechanical problem, namely,

$$\int_{\Omega_{-S}^0} \sigma_0 : \varepsilon_0(\mathbf{w}^0) \, d\Omega_{-S}^0(\mathbf{x}) + \int_{\Omega_S^0} \varepsilon_0(\mathbf{w}^0) : \mathbb{C}^0 \varepsilon_0(\mathbf{u}^0) \, d\Omega_S^0(\mathbf{x}) = \int_{\partial \Omega_{-S}^0} \hat{\mathbf{t}} \cdot \mathbf{w}^0 \, d\partial \Omega_{-S}^0(\mathbf{x})$$

$$\qquad (14)$$

being $\mathbb{C}^0$ obtained through a homogenization procedure (see ref. [17]).

   Note that Eq. (14) is a direct consequence of the linearity of the problem. In general or in nonlinear situations the separation between the macro and micro scales cannot be accomplished unless an appropriate linearization is conducted [36]. This statement is valid in classical multiscale linear mechanics where no evolution of the microstructure occurs. Nevertheless, since the considered microstructure is expected to grow up and to degrade, and additional hypothesis is introduced. We consider that both bone growth and scaffold degradation is averaged per time iteration. This implicitly considers that the macroscopic apparent properties are unchanged during the time step. A consequence of this simplification is that the time scales regarding to the application of loads and the diffusive analysis are uncoupled to the bone ingrowth and degradation times.

This multiscale model was implemented in an FE framework using the software Abaqus [37]. Parallel performance was used to accelerate the micromacro interaction (see [28] for details of the computer implementation).

3 Results: analysis of scaffold microstructural anisotropy

In the recent years, authors have drawn their attention to fabricate scaffolds with an internal anisotropic microstructure [38, 39]. This is usually referred as directional porosity which should guide bone regeneration inspired in actual bone structures such as a trabeculae. In this context, the presented multiscale approach is used in this work to analyze the effect of initial scaffold microstructural anisotropy on new bone tissue growth. To do this, a unit cell of several scaffold microstructures are studied as a representative volume element (RVE). Therefore, the corresponding microstructure of each macroscopic point of the scaffold specimen is featured by this RVE's. In order to reproduce scaffold microstructural anisotropy three cases are studied (see Fig. 4). The first two cases ((a) and (b) in Fig. 4) are characterized by an fcc arrangement of the empty pores [40]. In case (a) (Fig. 4a) pores are considered to be spherical with a porosity of 85%, whereas for case (b) (Fig. 4b) we consider that pores are elliptic with ratios $a/b = 1.5$ and $b/l = 0.31$ with 75% of porosity and oriented along the x-axis, being a,b elliptic semiaxis and l RVE size. Last case (c) (Fig. 4c) corresponds to the actual microstructure of a poly-ε-caprolactone (PCL) scaffold [17].

The degree of anisotropy of each scaffold microstructure may be evaluated by computing the macroscopic or apparent properties derived from its microstructure. Then, the fourth order stiffness tensors take the following values,

$$\mathbb{C}^0_{sph}(MPa) = \begin{pmatrix} 287.7 & 138.6 & 137.5 & 0 & 0 & 0 \\ & 284.7 & 139.3 & 0 & 0 & 0 \\ & & 287.6 & 0 & 0 & 0 \\ & sim & & 94.9 & 0 & 0 \\ & & & & 95.7 & \\ & & & & & 97.0 \end{pmatrix} \qquad (15)$$

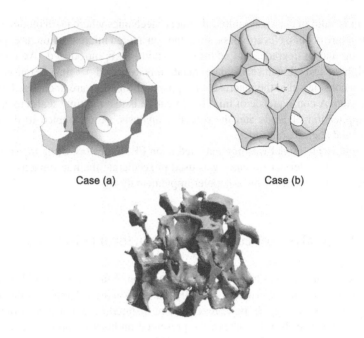

Case (a)   Case (b)

Fig. 4 Different cases of microstructure of scaffolds to study the effect of anisotropy. Case (a): fcc spherical scaffold 85% of porosity. Case (b): fcc elliptical scaffold oriented along x-axis 75% of porosity. Case (c): actual scaffold microstructure of 85±5 % of porosity [27].

$$
\mathbb{C}^0_{ell}(MPa) = \begin{pmatrix} 504.8 & 161.6 & 161.6 & 0 & 0 & 0 \\ & 329.4 & 162.1 & 0 & 0 & 0 \\ & & 329.4 & 0 & 0 & 0 \\ & sim & & 105.1 & 0 & 0 \\ & & & & 105.1 & 0 \\ & & & & & 94.1 \end{pmatrix} \tag{16}
$$

$$
\mathbb{C}^0_{ani}(MPa) = \begin{pmatrix} 40.9 & 18.7 & 23.9 & 4.9 & 5.2 & 4.0 \\ & 137.9 & 32.8 & 5.3 & 2.4 & 6.3 \\ & & 78.2 & 4.5 & 1.8 & 3.9 \\ & sim & & 16.9 & 17.2 & 1.1 \\ & & & & 17.2 & 4.1 \\ & & & & & 25.7 \end{pmatrix} \tag{17}
$$

On the other hand, the permeability matrices are as follows,

$$
\mathbf{K}_{sph}(mm^2) = \begin{pmatrix} 3.15 \cdot 10^{-3} & 0 & 0 \\ 0 & 3.15 \cdot 10^{-3} & 0 \\ 0 & 0 & 3.15 \cdot 10^{-3} \end{pmatrix} \tag{18}
$$

Table 3 Microscopic bone growth model parameters.

Parameter	Description	Value
m	Exponent which quantifies the importance of the number of cycles [-]	4.0
$\hat{\rho}$	Bone density [g/cc]	1.92
Ψ^*	Reference value for the mechanical stimulus at the tissue level [MPa/day]	12.5
w	Half-width of the dead zone [MPa/day]	3.125
N	Number of cycles considered in the load history [cycles/day]	24000
E_m	Young's modulus of the matured bone [MPa]	20000
v_m	Poisson's ratio of the matured bone [-]	0.3
T_m	Maturation time [days]	2
D_0	Scaling diffusion coefficient [mm^2/day]	400.0
k	Empirical rate constant for bone growth [μg/MPa]	$4 \cdot 10^{-5}$
M_c	Threshold mass for bone voxel formation [μg]	0.1

$$\mathbf{K}_{ell}(mm^2) = \begin{pmatrix} 2.84 \cdot 10^{-3} & 0 & 0 \\ 0 & 2.53 \cdot 10^{-3} & 0 \\ 0 & 0 & 2.53 \cdot 10^{-3} \end{pmatrix} \qquad (19)$$

$$\mathbf{K}_{ani}(mm^2) = \begin{pmatrix} 2.61 \cdot 10^{-3} & 0 & 0 \\ sim & 2.83 \cdot 10^{-3} & 0 \\ & & 2.69 \cdot 10^{-3} \end{pmatrix} \qquad (20)$$

where superscripts *sph*, *ell* and *ani* are used to refer the spherical pore scaffold (case (a) in Fig. 4), the elliptical pore scaffold (case (b) in Fig. 4) and the general anisotropic scaffold (case (c) in Fig. 4). Note that even though the scaffold has a lower porosity (for example the elliptical fcc pattern versus the spherical one), the elliptical scaffold microarchitecture shows a slightly lower permeability since the shape of the interconnection of pores is also more important than the permeability itself.

The RVE's of such microstructures were selected distinguishing between solid and fluid parts. As simplification we considered that the RVE's represents the microstructures of the entire biomaterial, such that, their microstructures may be reproduced by a periodic repetition of such belonging RVE. The scaffold microstructures were meshed using voxel-FEs for simplicity purposes [20, 28]. The model parameters used for this simulation are shown in Table 3 whereas the scaffold properties are highlighted in Table 4.

On the other hand, the multiscale model described above was mathematically implemented for an application example found in the literature [41]. A healthy left

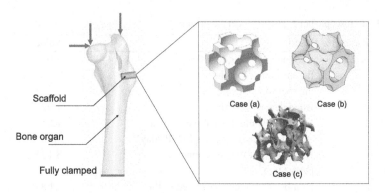

Fig. 5 Example of application in bone tissue engineering *in silico* using three different scaffold microstructures to evaluate the effect of anisotropy on new bone tissue regeneration [27].

Table 4 Scaffold parameters.

Parameter	Description	Value
E	Bulk biomaterial Young's modulus [MPa]	1570
v	Bulk biomaterial Poisson's ratio [-]	0.39
α	Hydrolysis model parameter [mm^2/day]	$4 \cdot 10^{-4}$
β	Hydrolysis model parameter [1/day]	4000
W_0	Bulk biomaterial molecular weight [Da]	114000
W_c	Molecular weight value under which the biomaterial degrades [Da]	10000

femur of a New Zealand rabbit was modelled by computer tomography (CT) using the Mimics software [42] (see Fig. 5). On this solid model, a virtual segmental 3×7 mm defect in the proximal third of the left femur, laterally to the greater trochanter [41], was performed in order to simulate the scaffold implantation. The FE mesh of that model was obtained using the Harpoon software [43] with four-noded bilinear tetrahedra both in the bone and scaffold macroscopic domains. Femoral loads were applied at the proximal part considering the rabbit hopping according to ref. [44]. Moreover, the initial femur density was obtained by applying the remodelling model presented above [30] in the healthy femur until convergence.

Note that the presented model is established over a strain energy-based rule as a stimulus for new bone tissue regeneration according to Eq. (6). Since we obtain the strain energy distribution at the microscopic level following a multiscale procedure (see refs. [17, 28]), it yields to an isotropic strain energy distribution when the scaffold has a symmetric pattern. Fig. 6 shows the strain energy distribution over the considered scaffold microstructures at the first step of analysis. In this figure, we can observe a symmetric chart of the energy strain distributions when the microstructure is considered to be an fcc spherical geometry (see Fig. 6a). On the other hand, Fig. 6b shows that the strain energy distribution is not symmetric due to the inclusion

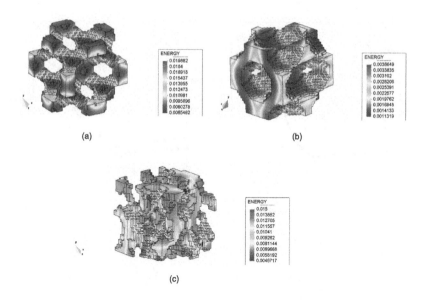

(a)   (b)

(c)

Fig. 6 Strain energy density distribution (MPa) over the considered scaffold microarchitectures. (a) Spherical fcc pattern, (b) elliptical fcc pattern and (c) fully anisotropic (actual) scaffold microarchitecture [27].

of elliptic pores along a certain direction of the fcc microarchitecture. Analogously, the full anisotropic scaffold microarchitecture (Fig. 6c) does not follow a symmetric pattern because of its anisotropic microstructure. Note that energy concentrations rise at certain locations due to the effect of changes of curvature.

The fact that symmetric scaffold patterns implies a symmetric distribution of the mechanical stimulus microscopically, yields to the conclusion that bone regeneration occurs volumetrically (symmetrically) induced by the scaffold microarchitecture. In Fig. 7 are plotted the evolution of the components C_{11}, C_{22} and C_{33} of the macroscopic stiffness tensor which gives an idea of the anisotropy of the internal microstructure. The trend of these curves are as follows: first we observe a decay of the mechanical properties due to hydrolysis until bone regeneration starts (see additionally Fig. 7), after this, mechanical properties increase due to the contribution of new tissue formed and finally mechanical properties degradation occurs due to the disappearance of the scaffold as consequence of the hydrolysis process. Interestingly, Fig. 7a shows that the evolution of the mechanical properties at x-direction (C_{11}), y-direction (C_{22}) and z-direction (C_{33}) are quite similar which means that new bone regeneration takes place symmetrically as consequence of the symmetry of its internal microstructure. On the other hand, when the scaffold microstructure has an induced anisotropy over a certain direction, bone regenerates preferently along that direction. This trend can be observed in Fig. 7b for the case of the elliptical fcc pattern. There, we can see that bone preferently appears along the direction of maximum stiffness, ie., x-axis (C_{11}) since the mechanical properties at this direction

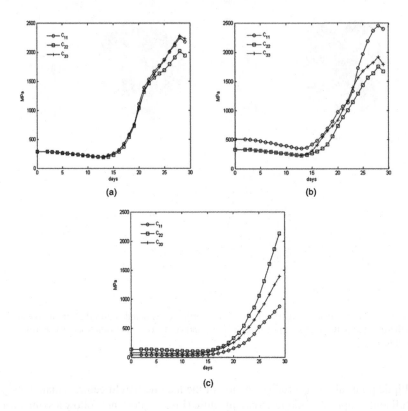

Fig. 7 Evolution of macroscopic (apparent) stiffness constants C_{11}, C_{22} and C_{33}. (a) Spherical fcc pattern, (b) elliptical fcc pattern and (c) fully anisotropic (actual) scaffold microarchitecture [27].

are the highest. Moreover, since this microarchitecture is symmetric over the other two directions, the evolution of the associated mechanical constants (C_{11} and C_{22}) are quite similar. Finally, the fully anisotropic scaffold microarchitecture yields to anisotropic formation of bone as consequence of the differences among mechanical constants at each direction in Fig. 7c.

The amount of new formed bone is obtained as the percentage between the new formed tissue volume versus the total RVE volume. It can be seen in Fig. 8, showing similar ratios of bone formation in the different scaffolds.

4 Conclusions

Scaffold parameters choice during design is a critical task for its subsequent success on a certain bone tissue engineering application. Some of the most important ones are focused on geometry (porosity and pore size) and biomaterial properties

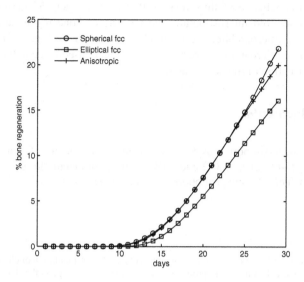

Fig. 8 Rate evolution of new bone tissue formation, obtained as the new formed tissue volume versus the total RVE volume, for the analyzed scaffold microarchitectures [27].

such as mechanical and degradation properties (in the case of biodegradable biomaterials). Porosity has been analyzed by authors concluding that the most important issue is to get an interconnected scaffold to allow nutrients exchange and waste removal [45, 46]. On the other hand, pore size effect was experimentally evaluated in [47, 48] showing higher rates of bone regeneration as pore size increases. Biomaterial physical properties are important as well in new tissue regeneration. In fact, stiffer biomaterials enhance bone tissue regeneration likely due to the mechanical stimulus sensed by the cells is more convenient and activates osteoblasts. Moreover, resorption kinetics should be slow enough to warrant scaffold integrity up to a bony structure has been formed [45]. A numerical study was performed in [16] analyzing the influence of each parameter on bone tissue regeneration.

On the other hand, two main conclusions may be derived from this work. The first one is that scaffold microstructure determines the subsequent architecture of the formed bone. In this work, this effect has been derived as consequence of the symmetry of the mechanical stimulus. Therefore, symmetric microstructures yield to symmetric patterns of bone formation. However, when anisotropy is present in the scaffold microarchitecture, bone regeneration tends to follow the direction of maximum stiffness. Therefore, using directional porosity we may modulate both the rate and internal microarchitecture of new bone tissue formation. Finally, in the case of a full anisotropic scaffold, bone regeneration is formed anisotropically although in the direction of maximum anisotropy as well, verifying therefore this hypothesis.

The second conclusion of this work, is that once we get an internal properly interconnected scaffold microstructure, the net rate of bone regeneration is very similar from one microarchitecture to another as proved by Fig. 8. However, as concluded before, microarchitecture has an important consequence in the shape and direction of formation of the new bone tissue.

Acknowledgements

This work has been supported by the Ministerio de Educacion y Ciencia of Spain (Project DPI2007-65601-C03-01) and CIBER-BBN Bioingeniería, Biomateriales y Nanomedicina. Their financial support is gratefully acknowledged.

References

1. Cao Y, Vacanti JP, Paige KT, Upton J, Vacanti CA (1997) Transplantation of chondrocites utilizing a polymercell construct to produce tissue-engineered cartilage in the shape of a human ear. Plast Reconstr Surg 100:297–304.
2. Kelly DJ, Prendergast PJ (2006) Prediction of the optimal mechanical properties for a scaffold used in osteochondral defect repair. Tissue Eng 12:2509–2519.
3. Kulig KM, Vacanti JP (2004) Hepatic tissue engineering. Transpl Immunol 12:303–310.
4. Ko HCH, Milthorpe BK, McFarland CD (2007) Engineering thick tissues–the vascularisation problem. Eur Cell Mater 14:1–18.
5. Langer R (2007) Tissue engineering: Perspectives, challenges, and future directions. Tissue Eng 13:1–2.
6. Shieh SJ, Vacanti JP (2005) State-of-the-art tissue engineering: From tissue engineering to organ building. Surgery 137:1–7.
7. Richardson TP, Peters MC, Ennett AB, Mooney DJ (2001a) Polymeric system for dual growth factor delivery. Nat Biotechnol 19:1029–1034.
8. Borenstein JT, Terai H, King KR, Weinberg EJ, Kaazempur-Mofrad MR, Vacanti JP (2002) Microfabrication technology for vascularized tissue engineering. Biomed Microdevices 4:167–175.
9. Warnke PH, Springer ING, Wiltfang J, Acil Y, Eufinger H, Wehmöller M, Russo PAJ, Bolte H, Sherry E, Behrens E, Terheyden H (2004) Growth and transplantation of a custom vascularised bone graft in a man. Lancet 364:766–770.
10. Cioffi M, Kuffer J, Strobel S, Dubini G, Martin I, Wendt D (2008) Computational evaluation of oxygen and shear stress distributions in 3D perfusion culture systems: Macro-scale and micro-structured models. Biomaterials 41:2918–2925.
11. Sanz-Herrera JA, Kasper C, van Griensven M, García-Aznar JM, Ochoa I, Doblaré M (2008a) Mechanical and flow characterization of Sponceram® carriers: evaluation by homogenization theory and experimental validation. J Biomed Mater Res B 87:42–48.
12. Hollister SJ, Maddox RD, Taboas JM (2002) Optimal design and fabrication of scaffolds to mimic tissue properties and satisfy biological constraints. Biomaterials 23:4095–4103.
13. Lin CY, Kikuchi N, Hollister SJ (2004) A novel method for biomaterial scaffold internal architecture design to match bone elastic properties with desired porosity. J Biomech 37: 623–636.
14. Taboas JM, Maddox RD, Krebsbach PH, Hollister SJ (2003) Indirect solid free form fabrication of local and global porous, biomimetic and composite 3D polymer-ceramic scaffolds. Biomaterials 24:181–194.
15. Gutierres M, Lopes MA, Hussain NS, Lemos AF, Ferreira JMF, Afonso A, Cabral AT, Almeida L, Santos JD (2008) Bone ingrowth in macroporous Bonelike for orthopaedic applications. Acta Biomater 4:370–377.

16. Sanz-Herrera JA, García-Aznar JM, Doblaré M (2009a) On scaffold designing for bone regeneration: A computational multiscale approach. Acta Biomater 5:219–229.
17. Sanz-Herrera JA, García-Aznar JM, Doblaré M (2009b) A mathematical approach to bone tissue engineering. Philos Transact A Math Phys Eng Sci 367:2055–2078.
18. Rajangam K, Behanna HA, Hui MJ, Han X, Hulvat JF, Lomasney JW, Stupp SI (2006) Heparin Binding Nanostructures to Promote Growth of Blood Vessels. Nano Letters 6:2086–2090.
19. Langer R (2006) Biomaterials for drug delivery and tissue engineering. MRS Bull 31:477–485.
20. Adachi T, Osako Y, Tanaka M, Hojo M, Hollister SJ (2006) Framework for optimal design of porous scaffold microstructure by computational simulation of bone regeneration. Biomaterials 27:3964–3972.
21. Wang Y, Pan J, Han X, Sinka C, Ding L (2008) A phenomenological model for the degradation of biodegradable polymers. Biomaterials 29:3393–3401.
22. Fauza DO (2003) Tissue engineering: Current state of clinical application. Curr Opin Pediatr 15:267–271.
23. MacNeil S (2007) Progress and opportunities for tissue-engineered skin. Nature 445:874–880.
24. Curtis A, Riehle M (2001) Tissue engineering: the biophysical background. Phys Med Biol 46:R47–R65.
25. Okuda T, Ioku K, Yonezawa I, Minagi H, Kawachi G, Gonda Y, Murayama H, Shibata Y, Minami S, Kamihira S, Kurosawa H, Ikeda T (2007) The effect of microstructure of -tricalcium phosphate on the metabolism of subsequently formed bone tissue. Biomaterials 28:2612–2621.
26. Stevens MM, George JH (2005) Exploring and engineering the cell surface interface. Science 310:1135–1138.
27. Sanz-Herrera JA, Doblaré M, García-Aznar JM (2010) Scaffold microarchitecture determines internal bone directional growth structure: a numerical study. J Biomech In press.
28. Sanz-Herrera JA, García-Aznar JM, Doblaré M (2008b) Micromacro numerical modelling of bone regeneration in tissue engineering. Comput Meth Appl M 197:3092–3107.
29. Einhorn TA, Majeska RJ, Rush EB, Levine PM, Horrowitz MC (1995) The expression of cytokine activity by fracture callus. J Bone Miner Res 10:1272–1281.
30. Beaupré GS, Orr TE, Carter DR (1990) An approach for time-dependent bone modelling and remodelling: theoretical development. J Orthop Res 8:651–661.
31. Doblaré M, García JM (2001) Application of an anisotropic bone-remodelling model based on a damage-repair theory to the analysis of the proximal femur before and after total hip replacement. J Biomech 34:1157–1170.
32. Doblaré M, Garcia JM (2002) Anisotropic bone remodelling model based on a continuum damage-repair theory. J Biomech 35:1–17.
33. Turner CH, Forwood MR, Rho J, Yoshikawa T (1994) Mechanical loading thresolds for lamellar and woven bone formation. J Bone Miner Res 9:87–97.
34. Tang L, Lin Z, Yong-Ming L (2006) Effects of different magnitudes of mechanical strain on osteoblasts in vitro. Biochem Biophys Res Comm 344:122–128.
35. Gopferich A (1997) Polymer bulk erosion. Macromolecules 30:2598–2604.
36. Terada K, Kikuchi N (2001) A class of general algorithms for multi-scale analyses of heterogeneous media. Comput Meth Appl M 190:5427–5464.
37. Hibbit, Karlsson, Sorensen. Abaqus user's manual v.6.2. HKS Inc. Pawtucket, RI 2001.
38. Lin AS, Barrows TH, Cartmell SH, Guldberg RE (2003) Microarchitectural and mechanical characterization of oriented porous polymer scaffolds. Biomaterials 24:481–489.
39. Mathieu LM, Mueller TL, Bourban PE, Pioletti DP, Muller R, Manson JAE (2006) Architecture and properties of anisotropic polymer composite scaffolds for bone tissue engineering. Biomaterials 27:905–916.
40. Brígido-Diego R, Más-Estellés J, Sanz JA, García-Aznar JM, Salmerón-Sánchez M (2007) Polymer scaffolds with interconnected spherical pores and controlled architecture for tissue engineering. Fabrication, mechanical properties and finite element modeling. J Biomed Mater Res B 81B:448-455.

41. Savarino L, Baldini N, Greco M, Capitani O, Pinna S, Valentini S, et al. (2007) The performance of poly-ε-caprolactone scaffolds in a rabbit femur model with and without autologous stromal cells and BMP4. Biomaterials 28:3101-3109.
42. Mimics Materialise. Mimics Materialise NV v10.0, 2006.
43. Harpoon Lt. Manchester: Sharc UK, 2006.
44. Gushue DL, Houck J, Lerner AM (2005) Rabbit knee joint biomechanics: motion analysis and modeling of forces during hopping. J Orthopaed Res 23:735-742.
45. Hutmacher DW (2000) Scaffolds in tissue engineering bone and cartilage. Biomaterials 21:2529-2543.
46. Karageorgiou V, Kaplan D (2005) Porosity of 3D biomaterial scaffolds and osteogenesis. Biomaterials 26:5474-5491.
47. Tsuruga E, Takita H, Itoh H, Wakisaka Y, Kuboki Y (1997) Pore size of porous hydroxyapatite as the cell-substratum controls BMP-induced osteogenesis. J Biochem (Tokyo) 12:317-324.
48. Kuboki Y, Jin Q, Takita H (2001) Geometry of carriers controlling phenotypic expression in BMP-induced osteogenesis and chondrogenesis. J Bone Joint Surg Am 83A:S105-S115.

Geometric modeling and analysis of bone micro–structures as a base for scaffold design

Y. Holdstein, L. Podshivalov, A. Fischer

Abstract Bones consist of hierarchical bio-composite materials arranged in multi-scale structural geometry. This structure is vulnerable to various damaging factors such as accidents, medical operations and diseases that may cause its degradation. Imaging techniques can already provide highly detailed micro-features of a bone or even its complete volumetric micro-structure. A three-dimensional model of the bone can then be reconstructed and analyzed. However, current technology cannot precisely fix damaged bone tissue and can only roughly approximate such damaged structures by using scaffolds with standard geometry. This chapter proposes a new method for creating natural scaffolds that can adapt according to location, size and shape. The method is based on constructing the scaffold as a 3D volumetric texture that imitates the irregular textural behavior of its surroundings. The method has the ability to create a smooth and continuous structure according to topological and geometrical characteristics. Moreover, the texture captures the stochastic and porous nature of the bone micro-structure. The resulting scaffold texture is tested by applying mechanical analysis to the new synthesized structure, thus controlling the mechanical properties of the reconstructed bone. We believe our method will help in customizing the design and fabrication of scaffolds for bone micro-structures. Moreover, such scaffolds can facilitate the process of rehabilitating damaged bone.

Y. Holdstein
Faculty of Mechanical Engineering, Technion – Israel Institute of Technology, Haifa 32000, Israel

L. Podshivalov
Faculty of Mechanical Engineering, Technion – Israel Institute of Technology, Haifa 32000, Israel

A. Fischer
Faculty of Mechanical Engineering, Technion – Israel Institute of Technology, Haifa 32000, Israel
email: meranath@technion.ac.il

P.R. Fernandes and P.J. Bártolo (eds.), *Advances on Modeling in Tissue Engineering*,
Computational Methods in Applied Sciences 20, DOI 10.1007/978-94-007-1254-6_6,
© Springer Science+Business Media B.V. 2011

1 Overview

Currently the biomedical community has shown major interest in creating and developing scaffolds to be used as a base for bone micro-implants. The following discusses the motivation for this interest, formulates the problem and proposes a new method for creating bone scaffolds for micro-structures. The scaffold structure dictates how the bone grows, and the material forming the scaffold is consumed by the bone. This material eventually diminishes over the years as new healthy bone replaces the scaffold material. The assumption is that the resulting bone structure will be influenced by the scaffold structure. In the bio-material field, recently research has focused on selecting biodegradable materials that can be absorbed by the bone, and on bio-manufacturing schemes for creating scaffolds that preserve the strength of the bone structure [1]. Since standard structures are characterized by a simple and symmetric layout, they are not customized to a specific patient, type of bone, or locality inside the bone (Fig. 1). Note that in these implants, only the size and density of the holes vary.

However, the structure of bone is complex at the micro-level. Bone is constructed from thin rods known as trabeculae and plates. These rods and plates are arranged in semi-regular 3D patterns and constitute highly anisotropic and heterogenic material (Fig. 2). This structure is stochastic in nature, and varies widely according to patient, bone type and location within a specific bone. The main issue with scaffolds to date is that little attention has been paid to geometrically modeling the structure of these scaffolds. Therefore, there is currently a gap between the optimal structure and the desired natural structure. We believe our work will contribute considerably towards closing this gap. The basic assumption underlying our work is that bone growing around a scaffold will acquire the scaffold shape faster if the scaffold has a more natural shape. Moreover, the resulting bone will integrate better with its surroundings and will thus perform functionally better than bone grown on a simple scaffold, especially at the interface between the scaffold and its neighborhood.

More specifically, our aim is to develop structural fillings, i.e., scaffolds that vary in location, size and shape. Such a filling may have the same irregular textural behavior as its surroundings. It should connect smoothly, according to topological and geometrical characteristics. Moreover, its mechanical properties should be topologically optimized with respect to the neighborhood of the hole. This study proposes a new approach for creating 3D micro-scaffolds for bone using 3D texture synthesis. Such an application has not yet been developed for irregular 3D micro-structures; thus, developing a new method poses a great challenge.

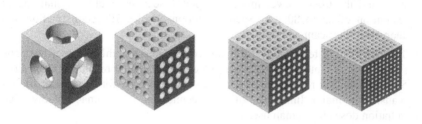

Fig. 1. Typical block structures of scaffolds used in state-of-the-art commercial systems [1].

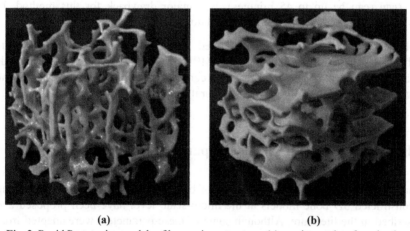

(a)   (b)

Fig. 2. Rapid Prototyping models of bone micro-structure: (a) specimen taken from lumbar spine; (b) specimen taken from femoral bone.

2 Bone micro-imaging technologies

Recent improvements in medical imaging technology provide high resolution *invivo* scanning of large specimens or even of whole bone models. These methods, all based on the proven technology of CT and MRI, are: (a) peripheral Quantitative Computed Tomography (pQCT); (b) micro Computed Tomography (μCT); and (c) micro Magnetic Resonance Imaging (μMRI). In the following, each method is described in detail.

(a) pQCT is a form of QCT that exposes the central body to a lower dose of radiation and is used for measuring the density of bones in limbs, such as the wrist and the tibia. Newer models of pQCT scanners yield nominal spatial resolution of up to 80 μm. Such resolution enables 3D reconstruction and analysis of the scanned micro-structure.

(b) The use of μCT technology for evaluation of micro-architecture of the bone started approximately a decade ago, when initial spatial resolution was about 50 μm. Today a resolution of 6 μm can be achieved for *in-vitro* or biopsy scans. The major limitation of this technology is its unacceptably high radiation doses for human use.

(c) μMRI technology has been developed in parallel with μCT technology, but only recently has become a prominent tool for bone micro-architecture evaluation. The absence of harmful ionizing radiation allows performance of high resolution *in-vivo* scans so that this technology is a potential future a candidate to replace the BMD test. However, the acquisition time of μMRI scan (can be up to 45 minutes), is a major drawback for this method. In addition, the resulting images are still noisy due to tissue interference.

Recently developed 3D computerized methods for reconstruction of the trabecular structure from micro CT and MRI images have enabled these methods to be used. Moreover, increasingly large computational resources have led to the development of analytical tools for structural and mechanical analysis.

3 Extracting 3D bone structural parameters

Quantitative and qualitative parameters of bone micro-architecture for analyzing reconstructed 3D bone models at the micro-scale level have been proposed and described in the literature. Although some of these parameters were adapted from analyses of 2D medical images, most are based on direct 3D analysis of volumetric models. Structural and topological parameters have been determined by commercial systems and in the literature. The following parameters have been calculated based on structural analysis of the 3D bone model [2,3]:

(a) *Basic parameters*: Bone Surface (BS) and Bone Volume (BV) are used for estimating the surface (BS) and volume (BV) of bone tissue. These parameters have characteristic values at each anatomic site and thus can be used to quantify bone porosity.

(b) *Trabecular architecture descriptors*: Trabecular Thickness (Tb.Th), Trabecular Separation (Tb.Sp.) and Trabecular Number (Tb.N) describe the trabecular architecture of the bone, which can vary as a function of anatomical site and as a consequence of disease between two distinctive structures: a rod-like structure and a plate-like structure. The parameters are used to quantitatively estimate trabecular thinning and to measure marrow cavity thickness and number of plates per unit length, respectively. Their values can be calculated directly from a 3D model or by using a set of the basic parameters, assuming plate-like architecture.

(c) *Advanced structural descriptors*, Marrow Star Volume (Ma.St.V), Structure Model Index (SMI) and Percent Plate (%Plate): Instead of trabecular separation, the Marrow Star Volume parameter (Ma.St.V) can be used to describe a measure of the "voids" within the trabecular structure that increase with progressive bone lost. The Ma.St.V is a sensitive descriptor for quantifying bone loss either through trabecular thinning or loss of entire trabecula. Alternatives for the trabecular number parameter are a Structure Model Index (SMI), which varies from 0-3 (an ideal plate structure and an ideal rod-like structure respectively) and Percent Plate (%Plate), which allows quantitative estimation of the effect of bone resorption on the shape of the trabecula.

(d) *Anisotropy descriptors*: Degree of Anisotropy (DA) and Percent Bone in Load Direction (%Bone) are used to describe anisotropy of the trabecular bone structure. The DA parameter is used for defining the direction of the preferred orientation of trabeculae and is important for directionally dependent mechanical properties. The second parameter allows quantitative estimation of the effect of bone resorption on the shape of the trabeculae.

The following parameters were calculated based on topological analysis of the 3D bone model [4,5]:

The connectivity descriptor: The Connectivity Density (ConnD) parameter provides an estimate of the number of trabecular connections per volume unit. It is defined as the number of trabecular elements that may be removed without separating the network and are frequently referenced as a parameter mostly affected during the progression of osteoporosis.

Process descriptors: Surface-to-Curve ratio (S/C) and Erosion Index (EI) parameters are part of a procedure called Virtual Bone Biopsy (VBB) and are used to provide detailed insight into the 3D trabecular network topology. The first is the surface-to-curve ratio (S/C), which is expected to be a sensitive indicator for the conversion of plates to rods. The second is the Erosion Index (EI), defined as the ratio of the sum of parameters expected to increase upon osteoclastic resorption (edges), divided by the sum of parameters expected to decrease due to such processes (surface).

These parameters and their combinations allow better understanding of "bone quality", and most have already been integrated into commercial applications provided with μCT and μMRI scanners. However, they yield only implicit and partial descriptions of bone topology, structure and geometric properties (shape elements and orientation). Therefore, a complete 3D micro representation is preferable.

4 Reconstruction of bone micro-structures

3D reconstruction from µCT/µMRI images produced by commercial systems often results in a highly distorted triangulated mesh that is not suitable for finite element mechanical analysis, and that therefore requires remeshing and mesh optimization. These operations are time consuming and sometimes demand manual user interventions. We examined the quality of 3D mesh models created from µCT images by a typical commercial scanning system. For the evaluation we used two characteristic parameters: (a) an aspect ratio that stands for a ratio between the longest and shortest edge of the triangles, ideally equal to one, and (b) maximum/minimum angles that in an ideal mesh are 60 degrees each. Tests showed 10%-50% failure rates. Therefore, new reconstruction methods are needed that eliminate the drawbacks described above. We have developed the following surface/volumetric reconstruction methods:

(a) Surface reconstruction from µCT/µMRI images, which are defined as parallel cross-sections, yielding a 3D triangulation model. A neural network method for surface reconstruction has been developed for macro and micro structures [6]. The resulting high quality surface model can be used for visualization and rapid prototyping (RP) processes.

(b) Volumetric reconstruction from the surface model, in which the basic volumetric unit can be either a hexahedron or a tetrahedron. This method has been developed for macro structures and is based on anisotropic grid-based technique [7].

5 Scaffold design using texture synthesis techniques

Texture synthesis is a sub-field of image processing, in which a picture whose size is theoretically unlimited is produced from a finite picture. All texture synthesis schemes produce a smooth and seamless picture that is larger than the original sample. These methods have the ability to recreate repetitive patterns appearing in the sample image, where the sample image is not usually able to be tiled. Heeger [8] proposed using a random noise image to generate texture according to the histograms of filter responses at different spatial scales. Another approach presented by De Bonet [9] uses multi-scale filter-based techniques. According to this approach, a texture patch is determined by its predecessors located in coarser scales. This scheme takes the input sample image and randomizes it so that the inner-scale dependencies are maintained.

The two main approaches to 2D texture synthesis are:

a. Pixel-by-pixel methods, where the pixels of the synthesized image are evaluated one at a time [10,11].

b. Block-by-block methods (patch based), where the synthesized image is created by fitting a cluster of pixels each time, such as the quilting algorithm [12].

In addition to its use for creating pictures, texture synthesis is used by two other applications:

• Image extrapolation, where an image is given without its borders and the boundary thus has to be approximated to agree with the image's internal content.

• Image in-filling, where a given image is mostly known but contains one or more regions where the values of the pixels are unknown. These holes must be approximated so that they fit the contents of the known neighborhoods. Moreover, the boundary between the old and the new regions must be smooth.

We examined the effect of block size on the resulting synthesized image. These results were generated by running the texture quilting algorithm (Fig. 3). Note that for a small block size, the result is more chaotic, and as the size grows the result becomes more and more repetitive. Thus, an optimal block size should be determined so that the synthesized image is not obviously repetitive, and its elements are not ruined. After applying various methods of texture synthesis, we made the following observations:

The quilting method [12] produces high quality images very rapidly. For example, the image shown in Fig. 3.b was built in less than a minute. The size of the block affects the nature of the synthesized image. As the block size gets smaller, the outcome is more chaotic and fewer features of the sample image are preserved. A pixel-by-pixel approach based on texture synthesis can be accelerated by implementing efficient image representation, as done by Wei & Levoy [11].

To measure and quantify how well the synthesized images fit their originating samples, we applied the proposed measurements (Fig. 3 and Fig. 4):

a. The correlation between the sample and the 2D synthesized images.

b. The correlation between an average pattern in the sample and the 2D synthesized images.

The resulting images of the quilting were analyzed by cross-correlating between the sample and the synthesized images. The intention is to determine how dominant the sample is with respect to the outcome of the algorithm. As shown in Fig. 3, the window size within the sample dramatically influences the distinguishable repetitiveness of the result (the white points on the second row of Fig. 3).

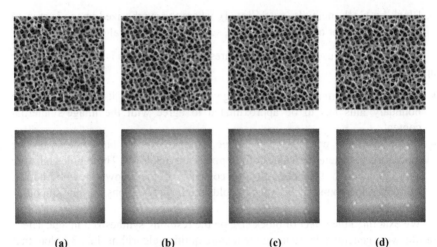

(a)   (b)   (c)   (d)

Fig. 3. Results from applying texture quilting on (a) sample image with size of 128x128 pixels; (b) 322x322 synthesized image from tile size of 10; (c) 348x348 synthesized image from tile size of 80; (d) 349x349 synthesized image from tile size of 100 and (f) 320x320 synthesized image from tile size of 120. The correlation result according to the proposed analysis is shown under each image.

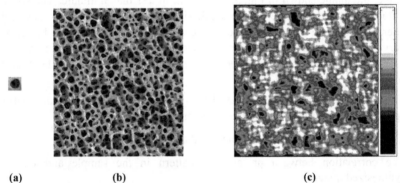

(a)   (b)   (c)

Fig. 4. Correlation results according to proposed analysis with (a) a manually selected averaged hole; (b) the resulting image; (c) the correlation result. The scale goes from dark to light for worst and best correlations, respectively.

Extension of 2D into 3D methods has the potential to assist in designing new micro-structures that can be used as the desired scaffold. This approach may guaranty the continuity of the resulting scaffold with its surrounding, while capturing the stochastic nature of the bone. Several attempts have been made to apply texture synthesis schemes to 3D models. However, these studies focus on synthesis of the 3D surface, rather than on the 3D volumetric structure. One notable study by Zhou et al. [13] proposes fitting a 3D structure to a given surface. The 3D structure is built by synthesizing a seed taken from a 3D sample structure.

This approach was motivated by the two-stage method introduced by Efros and Freeman [12]. First, the quilting of a 3D structure over a planar surface is detailed, and then, an extension is presented for applying the quilted mesh to a freeform surface. Although the results are of high quality, the method is not suitable for volumetric hole in-filling. Two other studies [14,15] also explore the issue of 3D geometric texture. Neither of these studies works directly with meshes, and both adopt a more restrictive representation. Bhat et al. [14] use a voxel-based approach, whereas Lagae et al. [15] use distance fields. Neither of these methods attempts to deal with 3D hole in-filling in the context of irregular patterns.

The previous sections discussed several texture synthesis methods. While some texture synthesis methods are faster than others, they are not necessarily suitable for the task of hole in-filling, since they may not have the ability to consider boundary conditions. A hybrid extension to 3D using both pixel-by-pixel and patch-wise approaches seems more appropriate for this purpose, since it has more degrees of freedom when choosing a new pixel value. Our method is inspired by the studies of [10,11,13]. Although not all have been extended to 3D, they have the potential to be applied in growing 3D irregular textures, given surrounding geometry and topology. To the best of our knowledge no existing method deals with micro-structures characterized by irregular texture.

6 The approach

The problem of designing and creating scaffolds for 3D volumetric cavities in bone micro-structure is still open and highly complex due to the irregular nature of bone micro-structure. A solution to the problem may provide powerful tools for physicians and medical developers in creating micro-scaffolds for bones at risk of fracture. One of the major innovations of our approach is that it works on irregular 3D volumetric bone textures, such as the trabecular bone micro-structure, rather than on segments and their topological relations. Our experiments so far have found that the texture synthesis approach is more natural for bone micro-structures. Following are the main stages of the proposed approach:

Scanning medical model – µCT/µMRI and extracting micro-structure from 2D/3D image: Initially, the medical data is acquired either from µCT or µMRI images, where the input is composed of digitized slice-by-slice 2D images. 2D diagnostic methods may be applied to these images. Alternatively, a 3D model may be constructed from 2D images [7] or cloud of points [6], allowing application of 3D diagnostic methods.

Extracting 3D holes of bone micro structure: This stage seeks to identify a volume in the mesh that represents a hole in the bone structure. This volume is characterized by sparse and relatively thin trabeculae.

3D hole in-filling, using Extended Voxel-by-Voxel (EVolV) texture synthesis: In this stage the holes are filled by using samples taken from a healthy region. Each hole is filled by a volumetric texture synthesis method. The input to the method is the cavity (hole) to be filled and the sample pattern. The holes are filled under the assumption that the sample for each hole can be extracted from the region around the hole of the damaged bone. This region is assumed to have an uncorrupted structure that defines the bone pattern. In addition, in certain cases where we may want to preserve some features of the original model, a feature extraction method for suggesting and assisting the growth process may be applied, as in [16].

FE mechanical analysis: This stage is important for understanding the structural properties of the synthesized material. The original 3D model and its in-filled region are subjected to 3D finite element analysis in order to discover high stress locations in the model and to detect discontinuation at interfaces boundaries.

Micro-structure based scaffold: One desired outcome of this work is its contribution to medicine. The resulting synthesized model may be manufactured utilizing existing micro-RP technologies. Such technologies have already acquired sufficient knowledge of bio-materials, allowing production of structures suitable for implantation in living tissues. Such structures are used as scaffolds for bone growth where needed, i.e. where there is injury.

7 The texture synthesis algorithms

Initially, the target image is synthesized of tiled blocks, randomly taken from the set of all overlapping blocks in the input texture image. Since there are noticeable boundary edges between neighboring blocks in the synthesized image, the next step is to add some overlap to the placement of the blocks when tiling them. In addition, instead of randomly selecting blocks, the process uses some measure to locate a tile that is correlated, at the overlap, with its neighbors that have already been placed. As the blocks are still likely to be distinguishable, an edge-smoothing process is applied between blocks by finding a minimum cost path through an error surface created in each overlap area. This is then set to be the new boundary between the (no longer) overlapping blocks. Although the size of the block is user-defined, it is intuitively determined. On the one hand, the block should be large enough to capture the relevant bone patterns in the texture. On the other, it must be small enough to prevent noticeable repetition in the resulting synthesized model. In addition, the size of the overlap area between adjacent blocks is pre-defined.

This section discusses the proposed extension to 3D in more detail. A straightforward 3D extension of 2D pixel-by-pixel does not suffice since connectivity issues occur due to the binary nature of the 3D model. Thus, the method was extended into a block-by-block based algorithm where the features were captured better and the connectivity was preserved. The advantages of the method are: a) The method produces good synthesized and seamless images; b) The method works very fast; and c) The user has to deal with only one parameter. The drawbacks of the method are: a) The method is not suitable for hole in-filling; and b) The method is non-deterministic.

7.1 The voxel-by-voxel algorithm

The proposed algorithm was initially based on the pixel-by-pixel approach introduced by Efros & Leung [10]. The following steps outline a 3D extension of the 2D case.

For each voxel in the synthesized mesh:
- Determine a cubic block size from the synthesized grid, centered at the current voxel.
- Determine a cubic block size from the sample mesh.
- Define a distance measure between the voxels of the sample and the corresponding ones from the synthesized grid.
- Create the set regions of the sample mesh that have a good correlation to the region from the synthesized grid.
- Randomly select a region from the set created in the previous stage and assign its center voxel value to the current voxel in the synthesized mesh.

7.2 The extended voxel-by-voxel (EVolV) algorithm

This section describes a further extension to the voxel-by-voxel method. The proposed EvolV algorithm is based on an amalgamation of the pixel-by-pixel approach [10] and patch-by-patch approach [12].
- The sample is scanned as in the voxel-by-voxel method.
- When a match is found in the sample, the entire block is copied into the synthesized model rather than only one voxel.
- The synthesized patterns may be further improved by integrating a mask that contains the main features in high contrast. This integration may be implemented by extending the weighting process for 3D patterns [16]. Creating the features mask may involve segmentation and feature detection.

As seen in the synthesized bone model, the result for the voxel-by-voxel method has some connectivity issues. These issues were diminished by utilizing the EVolV method. Figure 5 depicts results for four models: (a) The original model; b) model with artificially generated hole, (b) model with synthesized micro-structure and (c) model with symmetric scaffold where typical cavity has radius (R) of about 10 units ($\approx 350 \mu$m). The right column in Fig. 5 depicts a horizontal cross-section view of the respective models. These analysis results are discussed in detail in the following sub-section.

7.3 Calculating the principal direction of the anisotropic structure

For a given structure, directionality plays a major role in defining its mechanical properties. Thus, integrating such characteristic in the texture synthesis process is highly essential. Moreover, directionality within the bone's microstructure correlates highly with the loads exerted on the entire bone [17,18]. The trabeculae tend to be oriented according to the stress lines, maximizing the support the entire bone can provide. Since the orientation is a geometric property, it can straightforwardly be integrated in the texture synthesis process. Since this process is carried out through a greedy algorithm, the global feature of directionality might be corrupted. Thus, it is crucial to dictate that global feature throughout the entire process.

The first stage is to detect the directionality of the entire micro-structure in the region where missing bone has to be filled with a synthetic scaffold. Recall that the bone's microstructure is composed of irregular sub-structures and has a stochastic nature, so that this task is not trivial. For this purpose we initially applied the Fourier transform [19] on the bone model. This transform allows us to evaluate the frequencies and the latent energies depicting the structure. Afterwards, the transformed image is projected from different directions using Radon transform [20]. Then, the principal direction can be determined by extracting the projection with maximal energy. Finally the principal direction in the Fourier transform defines the direction of the pattern in the image [19]. Although the following example is shown on a 2D image for clarity purpose, the process was straightforwardly extended to 3D.

Figure 6 shows the results for two cases, where the directionality is different for each case. The scheme mentioned above successfully detected the principal directions in these textures. The first column presents the original image, the second column presents its Fourier domain, and the last column depicts the radon transform where the x-axis represents the angle of each direction. The marker indicates where the maximal value is reached.

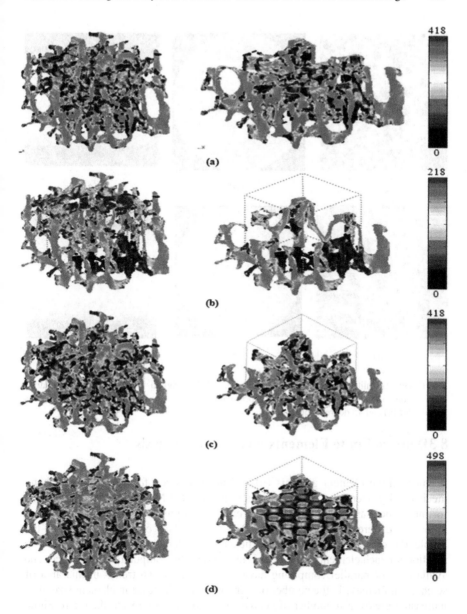

Fig. 5. EVolV texture synthesis and Micro FE analysis of: (a) original model, with (b) artificially generated hole, (c) the synthesized micro-structure and (d) symmetric scaffold where R is about 10 units ($\approx 350\mu m$). Maximal stress values on the scale are in kPa units.

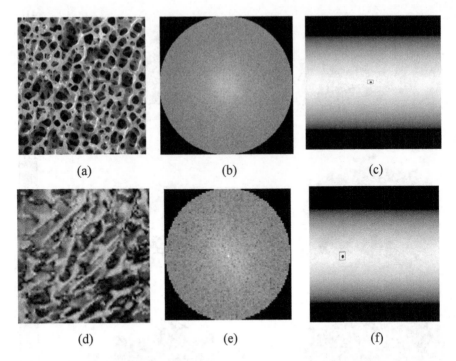

Fig. 6. Directionality assessment in textures. (a), (d) the original image; (b), (e) the Fourier domain and (c), (f) The radon transform with the maximal value marked with 86 degrees for (a) and 45 degrees for (d).

8 3D micro Finite Elements mechanical analysis

A micro finite element analysis (μFE) can be utilized for validating the bone mechanical properties. The finite element method (FEM) is one of the tools used for mechanical analysis. It is based on modeling a physical phenomenon as a set of partial differential equations (PDE) with boundary conditions, meshing the problem's domain with tetrahedral/hexahedral elements, and interpolating a solution for each element [21]. Today's more powerful computational resources and availability of parallel computing have contributed to two trends in this area of science: an increase in the number of elements and a decrease in element size, thus increasing the overall number of degrees of freedom. As a result, the micro Finite Element analysis technique [22,23] has been introduced. This technique provides a versatile tool for mechanical evaluation of bone tissue at the micro-structural level, thus facilitating prediction of possible fractures and damages. In most applications, hexahedral elements are preferable to tetrahedral elements for solving large FE applications, due to the efficient computational strategies developed for hexahedral elements [24]. These strategies are based on direct conversion of a voxel into a hexahedral element. In this case the voxel is degenerated to a "brick", thus avoiding the time-consuming computations of a stiffness matrix and a force vector for each element. State-of-the-art micro-FE methods today are based on parallel computing of high resolution large specimens of human bone reconstructed from μCT/μMRI images.

8.1 Bone tissue mechanical properties

At the tissue level, a bone's mechanical properties are strongly influenced by its micro-structure. Thus, a good scaffold should preserve the mechanical properties of healthy bone tissue. However, different structures can provide same mechanical properties; therefore, an additional parameter is required for evaluating scaffold quality. In the case of a trabecular structure, the porosity values can be considered to be a design parameter that influences the strength of the bone structure. The combination of these two parameters can lead to a topologically optimal structure. The mechanical properties of materials are characterized by the following set of parameters:

- E – Young's Modulus, describes the material's response to linear strain. Higher values of E represent a less elastic structure (higher rigidity).

- G – Shear modulus, describes the material's response to shearing strain.

- v – Poisson's Ratio, describes the ratio of the contraction strain orthogonal to the applied load, to the extension strain in the direction of the applied load.

Calculation of the above parameters at the tissue level of the bone is called a homogenization process. This process is commonly utilized for estimating the equivalent properties of composite materials, and bones are considered to be such materials.

For the equations of the mechanical properties of the bone tissue, two additional parameters are defined:

- σ – The structural stress, describes the average amount of force exerted per unit area.
- ε – The strain, describes the ratio of δ/L, where δ is the number of units of displacement of the boundary, and L is the initial dimension of the structure in the direction of the displacement.

The relationship between these parameters is defined according to Hooke's law:

- $\sigma = E\varepsilon$ –valid for stresses below the yield strength.

Applying a pre-defined displacement and proper boundary conditions to the model allows calculating the homogeneous parameters E, G and v for each material. The calculations are based on the average-strain theorem [25] that requires evaluation of the average values of stress and strain. The average stress over a volume V is defined as:

$$\bar{\sigma} = \frac{1}{V} \int_V \sigma dV \tag{1}$$

The average value of the strain $\bar{\varepsilon}$ is calculated from the displacement applied on the model:

$$\bar{\varepsilon} = \frac{1}{V} \int_V \varepsilon dV = \varepsilon^0 \tag{2}$$

Thus the homogenized material property E_{hom} at the tissue level can be calculated according to the Hooke's law as:

$$\frac{\bar{\sigma}}{\varepsilon^0} = E_{hom} \tag{3}$$

In this work we applied an iterative solution approach for mechanical analysis of a bone sample consisting of 300K hexahedral elements. Allowing three degrees of freedom for each element, we constructed a computational model with approximately one million unknowns. This procedure is part of an iterative solution based on the preconditioned conjugate gradient (PCG) method. The analyses were performed on a single PC Pentium dual-core 2.4 GHz with 4 GB of RAM. The analyses are based on the following evaluators:

The porosity of the resulting scaffold should be as close as possible to that of healthy bone micro structure. The design criteria should consist of optimal values of elasticity E and of the porosity *BV/TV*, so that the elasticity is not too low and the porosity is not too high with respect to healthy bone micro structure. In our case, $\delta = -1.28$ and $L=128$; thus the strain is 1% for all analyzed models, assuming a linear isotropic model [22]. Table 1 compares the mechanical properties of a healthy bone model, a damaged bone structure and different bone micro-scaffolds. The Young's modulus of the healthy bone and its BV/TV ratio are considered as a baseline and provide the design parameters for the scaffolds. The relative errors are shown with respect to this model. Four scaffolds are compared in this work. The first was designed using our proposed EVolV algorithm, and the other three were created using spherical voids with different radius. We applied the mechanical analysis on all these models, and the following conclusions may be drawn from the results:

- The scaffold designed by our approach satisfies both criteria of elasticity and porosity.
- Although synthetic structures may comply with the elasticity criteria as demonstrated in the case of R=7, it exceeds the optimal porosity value (about 50% error in average).

- The difference in elasticity values can be observed in the cross-section views of Fig. 5, according to the dominant dark colors which represent low stresses.
- In the case of the damaged model, it is clear that the remaining rods bear maximal stress, according to the dominant white and light grey coloration.

Table 1. A comparison between the mechanical properties of a healthy bone model, a damaged bone structure and different bone micro-scaffolds.

Bone Model	$\bar{\sigma}$ [kPa]	E [MPa]	E-Err [%]	BV/TV [%]	BV/TV Err. [%]
	Ave. Stress	Young Modl.		Porosity	
Healthy	111	11.1	–	10.3	–
Damaged	65	6.5	–41	5.9	–43
EVolV scaffold	111	11.1	0	10.2	1
Synthetic scaffold, R=5 units	88	8.8	–21	13.8	34
Synthetic scaffold, R=7 units	111	11.1	0	15.2	48
Synthetic scaffold, R=10 units	129	12.9	16	16.9	64

9 Summary and conclusions

The problem of designing and creating scaffolds for 3D volumetric cavities in bone micro-structure is still open and highly complex due to the irregular nature of bone micro-structure. A solution to the problem may provide powerful tools for physicians and medical developers in creating micro-scaffolds for bones at risk of fracture.

A novel method for patient-specific modeling of micro-implants for bones has been proposed in this study. The implants are constructed based on texture synthesis techniques. In 2D, μCT/μMRI images are used to generate a seamless image from a sample by applying the quilting algorithm. In this research, the 2D method is extended to a 3D method to achieve the desired outcome of volumetric in-filling of the micro-structure. These diseased cavities should be detected before the in-filling process is carried out. Such identification is not straightforward due to the porous and irregular nature of bone micro-structure. To the best of our knowledge, this is the first method to create 3D irregular texture for bone micro-structures. This method will facilitate customization in bio-manufacturing of micro-implants using micro-RP systems. The patent for the technology described in this paper is pending.

Acknowledgments

This study was partially supported by the Miriam and Aaron Gutwirth Memorial Fellowship, by the Phyllis and Joseph Gurwin Fund for Scientific Advancement at the Technion and by the Samuel and Anne Tolkowsky chair at the Technion.

References

1.   Almeida HA, Bartolo PJ, Ferreira JC. Mechanical behavior and vascularisation of tissue engineering scaffolds. In: Advanced Research in Virtual and Rapid Prototyping, Leiria: Taylor & Francis; 2008, p. 71-81.
2.   Feldkamp LA, Goldstein SA, Parfitt AM, Jesion G, Kleerekoper K. The direct examination of three-dimensional bone architecture *in vitro* by computed tomography. Journal of Bone and Mineral Research 1989; 4, p. 3-11.
3.   Hildebrand T, Laib A, Muller R, Dequeker J, Ruegsegger P. Direct three-dimensional morphometric analysis of Human Cancellous Bone: Microstructural Data from Spine, Femur, Iliac Crest and Calcaneus. Journal of Bone and Mineral Research 1999; 14, p. 1167-74.
4.   Odgaard A, Gundersen HJ. Quantification of connectivity in Cancellous bone, with special emphasis on 3-D reconstructions. Bone 1993; 14, p. 173-82.
5.   Gomberg BR, Saha PK, Song HK, Hwang SN, Wehrli FW. Topological analysis of trabecular bone MR images. IEEE Transactions on Medical Imaging 2000; 19, p. 166-74.
6.   Holdstein Y, Fischer A. Three-dimensional surface reconstruction using meshing growing neural gas (MGNG). Vis. Comput. 2008; 24, p. 295-302.
7.   Azernikov S, Fischer A. A New Approach to Reverse Engineering based on Volume Warping. ASME Transactions, Journal of Computing & Information Science in Engineering (JCISE) 2006; 6:355-363.
8.   Heeger DJ, Bergen JR. Pyramid-based texture analysis/synthesis. In: SIGGRAPH '95, Los Angeles: ACM Press; 1995, p. 229-238.
9.   De-Bonet JS. Multiresolution sampling procedure for analysis and synthesis of texture images. In: SIGGRAPH '97, Los Angeles: ACM Press; 1997, p. 361-368.
10. Efros AA, Leung TK. Texture Synthesis by Non-parametric Sampling. In: ICCV'99; 1999, p. 1033-8.
11. Wei LY, Levoy M. Fast texture synthesis using tree-structured vector quantization. In: SIGGRAPH '00, New Orleans: ACM Press; 2000, p. 479-488.
12. Efros AA, Freeman WT. Image Quilting for Texture Synthesis and Transfer. In: SIGGRAPH '01, Los Angeles: ACM Press; 2001, p. 341-6.
13. Zhou K, Huang X, Wang X, Tong Y, Desbrun M., Guo B, et al. Mesh quilting for geometric texture synthesis. In: SIGGRAPH '06, Boston: ACM Press; 2006, p. 690-7.
14. Bhat P, Ingram S, Turk G. Geometric texture synthesis by example. In: SIGGRAPH 2nd Symposium on Geometry Processing, Nice: ACM Press; 2004, p. 41-4.
15. Lagae A, Dumont O, Dutre P. Geometry Synthesis by Example. In: SMI'05: Proceedings of the International Conference on Shape Modeling and Applications, Boston: IEEE Computer Society; 2005, p. 176-185.
16. Lefebvre S, Hoppe H. Appearance-space texture synthesis. In: SIGGRAPH'06, Boston: ACM Press; 2006, p. 541-8.

17. Wolff J. The Law of Bone Remodelling. Berlin/New York: Springer-Verlag 1986; pp. 126.
18. Cowin SC. Bone Mechanics Handbook 2nd Edition. CRC Press, Boca Raton; 2001.
19. Gonzalez RC, Woods RE. 2006 Digital Image Processing. Prentice-Hall, Inc.
20. Deans, SR. The Radon Transform and Some of Its Applications. New York: John Wiley & Sons; 1983.
21. Podshivalov L, Holdstein Y, Fischer A, Bar-Yoseph PZ. 2D Multi-Scale Finite Elements Analysis as a Base for a 3D Computerized Bone Diagnostic System. International Journal of Communications in Numerical Methods in Engineering 2009; 25, p. 733-49.
22. van Rietbergen B, Weinans H, Huiskes R, Odgaardt A. A new method to determine trabecular bone elastic properties and loading using micromechanical finite-element models., Journal of Biomechanics, 1995, Vol. 28(1), p. 69-81.
23. van Rietbergen B. Micro-FE analyses of bone: state of the art. Adv. Exp. Med. Biol. 2001; 496:21-30.
24. Arbenz P, van Lenthe GH, Mennel U, Muller R, Sala M. Multi-level μ-Finite Element Analysis for Human Bone Structure. Workshop on State-of-the-art in Scientific and Parallel Computing 2006; Umeå, Sweden.
25. Qu J, Cherkaoui M. Fundamentals of micromechanics of solids. Wiley, New-Jersey, 2006, p. 108-9.

Electrospinning and Tissue Engineering

Geoffrey R. Mitchell and Fred Davis

Abstract We review how the process of generating scaffolds for tissue engineering using the process of electrospinning sits alongside other more established procedures for the preparation of scaffolds. We identify the key potential advantages of electrospinning for constructing scaffolds and we explore some of the remaining challenges. Several of these focus on the solvent and the desirability of using water for biological systems rather than highly volatile solvents such as those based on fluoroalcohols which lead, for example to the denaturing of collagen. We show how the control of temperature opens several opportunities as does the inclusion of additives to influence conductivity and viscosity. We show how the use of alternative electrode arrangements can lead to a control of the internal organisation with potential for property enhancement. It is reasonable to currently see electrospinning as an area dominated by experimental work but there is a growing computational element and we review how far such developments have reached and the potential for identifying new solutions for the application of electrospinning to tissue engineering.

1 Introduction

Electrospinning involves the generation of fine polymer fibres by means of applying an electric field to droplets of a polymer solution passed from tip of a fine orifice [1]. Under the direct influence of the Coulombic electric forces, a small droplet made up from a solution of a polymer will become charged; as a consequence its shape will become distorted to form a cone-like geometry [2]. Dependent on the strength of the applied electric field together with the solution viscosity, surface tension and dielectric properties of the solution, the integrity of the droplet will break down and a fine fibre of polymer will be produced and

Geoffrey R Mitchell
University of Reading, Whiteknights, Reading RG6 6AF UK
Polytechnic Institute of Leiria - Centre for Rapid and. Sustainable Product Development.
Campus 5. Rua das Olhalvas. 2414-016 Leiria. Portugal
email: g.r.mitchell@reading.ac.uk

Fred Davis
University of Reading, Whiteknights, Reading RG6 6AF UK

P.R. Fernandes and P.J. Bártolo (eds.), *Advances on Modeling in Tissue Engineering*,
Computational Methods in Applied Sciences 20, DOI 10.1007/978-94-007-1254-6_7,
© Springer Science+Business Media B.V. 2011

deposited on the collector electrode. During the flight, the fibre will be subjected to instabilities which extend the fibre and lead to a considerable reduction in the diameter [3, 4]. An important feature of this technique is that it produces fibres with diameters ranging from 10 nm to a few microns, dimensions not generally available using other techniques. The history of electrospinning is long and dates back over a hundred years [5, 6]. Recent interest in this technique stems from both the topical nature of nanoscale material fabrication, and the considerable potential for use of these nanoscale fibres in a range of applications including, amongst others, a range of biomedical applications processes such as drug delivery and the use of scaffolds to provide a framework for tissue regeneration in both soft and hard tissue applications systems [7].

In this chapter we first review how the process of generating scaffolds for tissue engineering using the process of electrospinning sits alongside other more established procedures for the preparation of scaffolds. We examine the challenges which still remain in the area of electrospinning and the state of advancement of modelling techniques especially as relate to biomedical applications.

2 Tissue Engineering

Tissue engineering sets out to apply the key principles of both engineering and life sciences to deliver biological substitutes that restore tissue function or in some situations generate whole organs. Key advances in biological materials especially in the area of stem cells, growth and differentiation factors generate realistic opportunities to create tissues in the laboratory using an engineered extracellular matrix or scaffold and biologically active molecules [8]. Here we focus on the requirements of the scaffold. This provides a biodegradable 3-d scaffold which can be implanted to aid the repair of tissue. The scaffold acts as an artificial extracellular matrix and it needs to mimic the chemical composition and physical architecture of natural extracellular matrix to facilitate cell adhesion, proliferation, differentiation and new tissue formation [9]. This places considerable demands on the physical and chemical properties of the material used, on the structure and morphology of the scaffold itself and on the active ingredients which can be incorporated in to the scaffold to aid tissue formation.

There are a variety of viable manufacturing approaches to producing a biodegradable 3-d scaffolds and the technique of electrospinning should be seen in that context. Key parameters for the structure and morphology of the scaffold include pore size and porosity as well as the critical factors of cell and tissue compatibility.

We will now review some of the principal scaffold fabrication techniques relevant to polymer based materials and compare these to the electrospinning methodology.

Direct Writing

Direct writing of scaffolds involves a layer-by-layer process which allows the fabrication of involved and especially complex architectures [10]. Such structures can be prepared at high resolution. As a consequence, the procedure of direct writing provides a successful approach to the accurate control of the pore size and the interconnectivity. There are three notable disadvantages, the first is that technique, akin to the desktop printer, requires relatively expensive equipment, the second is that only biomaterials which can be formed in to polyelectrolyte inks can be utilise and thirdly, the very nature of the layer-by-layer process means that the generation of scaffolds of the required thickness will take a substantial time.

Fused Deposition Modelling

Fused deposition of scaffolds also involves a layer-by-layer process which facilitates the fabrication of complex architectures. The disadvantage is that the technique is limited to polymer materials which melt with the attendant issue of degradation in the melt phase. This method has been applied to the generation of scaffolds for the trabecular bone using polybutylene terephthalate where the authors were able to reproduce the level of porosity exhibited in the natural bones [11].

Selective Laser Sintering

Laser sintering is a relatively expensive process and the choice of materials is even more limited that the fused deposition approach. The available materials are largely ceramic based but polymers have also been explored [12]. As with the previous techniques, the fabrication process facilitates excellent control of the pore size and the preparation of systems with well defined interconnectivity.

Stereolithography

Stereolithography is a layer-by-layer additive manufacturing process. As such it provides good control of macroscopic pore size as well as the ability to construct complex architectures. The major drawback to stereolithography in the current context is that it is restricted to UV curable materials. As the process involves chemical reactions, there are likely to be specific issues due to the presence of toxic by-products. Realistic progress in this area will require new types of UV curable resins as exemplified by the work of Melchels et al [13]. The technique also uses relatively expensive computer directed equipment.

Dimensional Printing

In principle three dimensional printing in a highly attractive fabrication approach but it does require the use of relatively expensive equipment and the materials need to be available in a controlled powder size. The latter may be difficult to achieve with many materials. However, the method does generate scaffolds with an excellent control of the pore size and architecture. The process has been used to prepare hydroxyapatite scaffolds with very specific architectures [14].

Self Assembling Nanofibres

Self-assembling scaffolds based on nanofibres can be prepared in water and offer the advantage that cells can be introduced in to the system during preparation. The technique generates low volumes of materials and the final product will generally have poor mechanical properties and hence the use will be restricted to soft tissues [15].

Thermally Induced Phase Separation

Generating scaffolds through thermally induced phase separation is simple process which can be used by a variety of biomaterials. It requires nothing other than general laboratory equipment and by adjusting the nature of the phase separation large scale porosity can be achieved. The technique usually employs organic solvents which may limit the materials and induce some toxity effects. The approach can involve more than one polymer and this can lead to substantial improvements in the attachment and growth of cells to the scaffold [16].

Particulate Leaching

Particulate leaching is another relatively simple approach but which again often employs organic solvents and the nature of the process means that it may be difficult to achieve a high interconnectivity of pores. It can be employed with a variety of materials and Reignier and Huneault [17] have extended the technique to involve the leaching of both polymer and salt components.

Electrospinning

Electrospinning is a relatively inexpensive method which can be applied to a very wide range of materials. The scaffolds currently produced often exhibit poor mechanical properties. A further disadvantage is that it may be difficult to achieve large pore sizes. The process of electrospinning involves the use of high dielectric constant polar solvents which means that in many cases the state of the biological material during the preparation of the scaffold may well be compromised. This is considered in section 4.

Reviewing electrospining alongside other scaffold fabrication processes reveals some significant advantages include the wide-range of materials available and the potential for eliminating the use of toxic elements during the fabrication. The advantages can be readily explored but there remains the challenge of building in a range of large pore sizes. Currently it would appear that in order to overcome this limitation electrospinning will need to be used in conjunction with another of the fabrication processes, for example an additive fabrication to yield the larger pores, while the fibres provide a finer scale structure [18].

3 Electrospinning

Electrospinning is deceptively simple in terms of the components as shown in Figure1. A working system can be readily assembled in a standard science laboratory, although some fabrication procedures will require more complex configurations. Commercial electrospinning systems are now becoming available. Typically, a high constant electric field (0.25 to 2 kV/cm) is applied between a needle point capillary tip and a grounded collection screen using a variable high voltage supply in the range 5-20kV. A glass syringe fitted with a metal needle provides a very convenient system for both holding and feeding a constant stream of fluid to the needle tip. The rate of fluid flow is usually controlled using a syringe pump and this facilitates continuous spinning over a prolonged period. The arrangement outlined above also provides a straightforward insulating support for the needle. For these reasons, this type of configuration appears in many laboratory based studies. Transfer of the technology to a production situation may involve modification to achieve volume throughput and reproducibility.

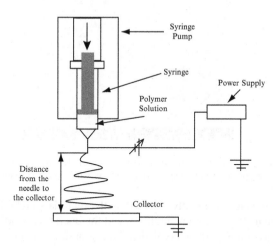

Fig. 1. A schematic of the key components in the electrospinning technique

The high voltage induces charge on the surface droplet at the end of the syringe capillary. The induced charge introduces mutual charge repulsion at the droplet surface creating a force which opposes the surface tension of the liquid. As a consequence, the liquid drop at the tip of the capillary tube elongates to form a progressively conical shape, often referred to as the Taylor cone. Above a critical electric field a fine jet is ejected from the tip of the Taylor cone which has a flight trajectory towards the collector electrode. During the flight, electrically induced bending instabilities result in hyper-stretching and the high surface area to volume ratio leads to rapid loss of solvent. These processes lead to the production of fibres ranging in diameter from tens of nanometres to a few microns. The process of transforming the solution in to polymer fibres during the flight is complex. The flight velocity is ~ m/s and the fibres are generally deposited as a random matt as shown in Figure 2. The cross-sectional shape of the fibres in this random matt is essentially circular but other shapes may be observed as highlighted in Section 4.1.

The electro-spinning process may be broken down in the following stages:

- Charging of the fluid
- Formation of the cone-jet
- Thinning of the jet in the electric field
- Instability of the jet
- Collection of the jet or its solidified fibres

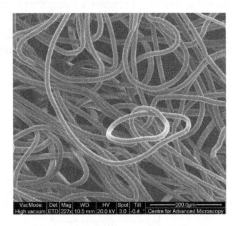

Fig. 2. A scanning electron micrograph of random matt of electrospun poly(ε-caprolactone) (Mw=80,000) fibres prepared from a 30% w/v dichloroethane solution.

Each of these stages is greatly affected by the materials including the polymer, the solvent system as well as any additives and the operating conditions employed. Clearly, there are many parameters which require optimisation to yield fibres with the correct scale and properties. The availability of these parameters is in part the source of the flexibility of the electrospinning technique. This means that it is possible to prepare electrospun fibres from virtually any polymer and indeed this is the case. There is now a considerable body of relevant literature. At the time of writing the Thomson-Reuter Web of Science returned over 5,000 scientific papers concerned with electrospinning and electrospun fibres.

There are clear material requirements for the preparation of scaffolds for tissue engineering applications in terms of toxicity, cell compatibility and biodegradability. The electrospinning process itself does not significantly limit the choice of material. To date, the overwhelming majority of studies have spun fibres from solution, although melt electrospinning is also possible [19]. Clearly, the solvent may itself influence the state of any biological polymers used. For example, it has shown that the use of fluoroalcohols such as 1,1,1,3,3,3-hexafluoro-2-propanol or 2,2,2-trifluoroethanol as solvents to prepare electrospun fibres of collagen leads to effective denaturing of the collagen. The fibres produced are effectively gelatine fibres [20, 21]. The use of toxic solvents in the manufacturing may, in any case, cause concern at the point of use although it is the case that the vast majority of the solvent is lost during the electrospinning process. In the example shown in Figure 2, the poly(ε-caprolactone) fibres were spun from a solution based on dichloroethane. NMR studies of as spun fibres dissolved in a deuterated solvent (CDCl$_3$) showed that only traces of the level of solvent remained [22]. If the fibres were left in air from 24 hours and then studied, no solvent could be detected. We explore the role of the solvent in the Section 4.

4 Solvents

4.1 Solvent type

The solvent is by far the largest component in the starting solution and therefore it has a major influence on almost every aspect of the electrospinning process. The solvent will have strong impact on the state of the polymer chains in solution; for example most organic solvents used in electrospinning are highly polar. In general, they will be some way from the classic ideal or theta solvent of polymer theory. Table 1 shows a number of common solvents together with their electrical properties and the boiling point.

The table of solvents shows a considerable variation in properties in just this small selection. For many polymers, the dielectric polarisation is low and it is the solvent that largely determines the force on the droplet via the interaction with the electric field through the dielectric constant and the conductivity. Figure 3 shows the cross-sections for electrospun fibres of atactic polystyrene prepared from dimethyl ether ketone (Figure 3a) while Figure 3b shows fibres spun under the same conditions but from a solution based on dimethylformamide. The latter fibres exhibit circular cross-sections but more detailed investigations reveal high levels of nano-porosity [23, 24]. The fibres spun from methyl ether ketone exhibit a dumbbell type cross section. These differences are attributed to the differences in volatility of the solvents. Methyl ether ketone exhibits a boiling point some 70 degrees below that exhibited by dimethylformamide. The dumbbells cross-section is thought to arise when a skin forms on a fibre in flight before all the solvent has evaporated. Subsequent evaporation of the solvent with the skin leads to a collapse of the outer skin. The presence of the solvent inside the fibre leads to additional forces which given rise to the dumbell ends [25].

Table 1. Typical solvents used in electrospinning.
The electric properties relate to room temperature

Solvent	Solvent Type	Boiling Point (°C)	Dielectric Constant	Dipole Moment D
Hexane	Non-polar	69	2.0	0.0
Toluene	Non-polar	111	2.4	0.36
Cholorform	Non-polar	61	4.8	1.0
Dichloromethane	Polar Aprotic	40	9.1	1.6
Tetrahydrofuran	Polar Aprotic	66	7.5	1.8
Methyl Ether Ketone	Polar Aprotic	80	18.5	2.8
Dimethyl formamide	Polar Aprotic	153	38	3.8
Acetonitrile	Polar Aprotic	82	37	3.9
Formic acid	Polar Protic	101	58	1.4
Ethanol	Polar Protic	79	30	1.69
Acetic Acid	Polar Protic	118	6.2	1.74
Water	Polar Protic	100	80	1.85

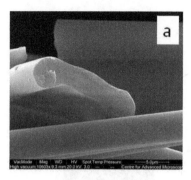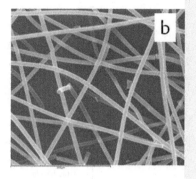

Fig. 3. Scanning electron micrographs of (a) atactic polystyrene fibres electrospun from a methyl ether ketone based solution and (b).the same material electrospun from a dimethylformamide based solution.

4.2 Solution Concentration

Probably the most critical operational parameter in an electrospinning system is the concentration of the polymer in the solution. At low concentrations of the polymer, the application of the electric field results in electrospraying in which the electrode becomes covered in small drops of polymer. Increasing the concentration of the polymer leads to the formation of fibres along with some beading as shown in Figure 4. For concentrations above a critical value, smooth continuous non-beaded fibres are obtained. Experimentally, increasing the molecular weight at a fixed concentration yields similar results. The critical level of concentration required has been explored by a number of authors for example, Shenoy et al. [26] studied fibres prepared using different polymers and concluded that stable fibre formation took place from solutions with an entanglement count $n_c > 2.5$ entanglements per chain. They proposed that n_c can be derived using equation 1.

$$n_c = \frac{\phi_p M_w}{M_e} \tag{1}$$

Where ϕ_p is the fraction of polymer in the solution M_w is the molecular weight of the polymer and M_e is the average molecular weight between entanglements. This model provides a particularly useful framework for understanding the variation of the critical concentration for different polymers and solvents.

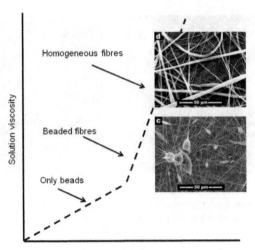

Fig. 4. A schematic of the relationship between solution viscosity and the concentration of the polymer solution. The scanning electron micrographs are for fibres of poly(ε-caprolactone) prepared from different concentrations.

At low concentrations ie in the dilute regime, there are no overlaps between random coils. As the concentration increases, overlap occurs and entanglements can form. At this point the viscosity versus concentration curves changes slope as shown in the schematic.

4.3 Solution Viscosity

The previous section has shown the importance of the solution viscosity in determining the outcome of the electrospinning procedure. Kariduraganavar et al. [27] have reported a novel approach to controlling the solution viscosity without adjusting the polymer concentration. They employed a low molar additive, dibenzylidene sorbitol (DBS), which is widely used as a gelator. The addition of small quantities to the solution based on dichloroethane and methanol led to a marked increase in viscosity as shown in Figure 5. In contrast, similar quantities in a solution based on dichloromethane and dimethylformamide resulted in a modest increase. Fibres electrospun from these two solvent systems showed remarkable changes in their diameter and morphology as shown in Figure 6. In the case of the dichloroethane – methanol system the addition of 1% of DBS is sufficient to inhibit the formation of beads even though the polymer concentration remains the same.

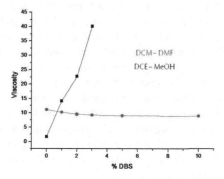

Fig. 5 A plot of the viscosity (mPa.s) of solutions of poly(ε-caprolactone) based on either dichloromethane/dimethylformamide or dichloroethane/methanol as a function of the additive dibenzylidene sorbitol

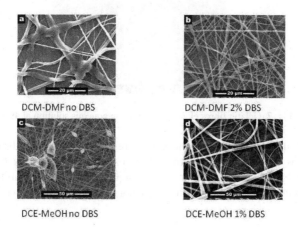

Fig. 6. PCL fibres prepared by electrospinning the solution systems shown for each micrograph.

4.4 Solvent Toxicity

Collagen and gelatin offer considerable promise as scaffold materials with regard to biocompatibility. Gelatin is a complex biopolymer system containing principally glycine, proline and 4-hydroxyproline. In aqueous solution it exhibits a gel at temperatures below ~ 30°C. The gelation is driven by hydrogen bonding and the formation of short sequences of triple helix structures similar to that observed in collagen. The glass transition of gelatin is strongly dependent on the water content; with 20% water Tg ~ 50°C. To successfully electrospin fibres from a particular solution, requires a solution concentration above the critical concentration for entanglements, essentially the solution needs to exhibit elasticity. At such concentrations, gelatin readily gels at room temperature and hence electrospinning is not possible. A number of studies have used non-aqueous acidic solvents including 2,2,2-trifluoroethanol, glacial acetic acid and acetic acid/dimethyl sulfoxide each of which destabilises the triple helix formation. These are aggressive solvents for gelatin and related proteins such as collagen and as shown by Zeugolis et al. [21] causes irreversible changes. There is considerable advantage of being able to make nanofibres from an aqueous solution. Water is non-toxic and it is not aggressive with respect to the gelatin. Elliott et al. [28] have explored an alternative approach by carrying out the electrospinning at an elevated temperature so that the gelatin solution is above the gel-sol point. Figure 7 shows electrospun gelatin fibres spun from a 30% w/v gelatin in water solution in which the syringe and the environment is held at 49°C.

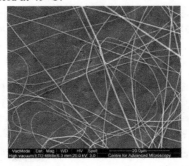

Fig. 7. Gelatin fibres electrospun from an aqueous solution at 49°C

Elliott et al. [28] show that nanoscale fibres of gelatin can be directly prepared from aqueous solution using thermal electrospinning techniques for a range of temperature >36°C. Fibres prepared from aqueous solution exhibit a mean diameter which increases slightly with increasing temperature as shown in Figure 8.

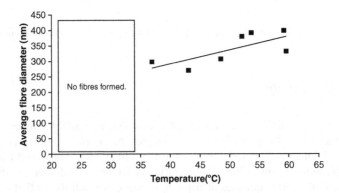

Fig. 8. The mean fibre diameter for gelatin fibres electrospun from 30% aqueous solutions plotted as a function of the temperature of the solution temperature.

Elliott et al. [28] used wide-angle x-ray scattering techniques to explore the presence or otherwise of triple helical sequences in the electrospun gelatine fibres. Figure 9a and 9d contrast the scattering obtained for bulk gelatin samples prepared at 20°C and at 60°C. In the former, sharp and distinct peaks can be observed at $|Q|{\sim}0.5\text{Å}^{-1}$ corresponding to the correlations between helices and at $|Q|{\sim}2.2\text{Å}^{-1}$ corresponding to the correlations within the helical sequence. $|Q| = 4\pi\sin\theta/\lambda$, where λ is the incident wavelength and 2θ is the scattering angle. The scattering from gelatin fibres prepared by electrospinning at 40°C (Figue 9b) shows no evidence for triple helical sequences as was expected. The authors attribute this to the limited time at which there was sufficient water in the fibres to provide mobility for the formation of the ordered sequence during the electrospinning flight. Fibres exposed to water vapour for several hours at room temperature lead to the re-appearance of the characteristic diffraction peaks (Figure 9c) confirming that the high temperature electrospinning had not degraded this particular aspect of the gelatin structure.

The success of the electrospinning at elevated temperatures suggests that this is an area which merits further study.

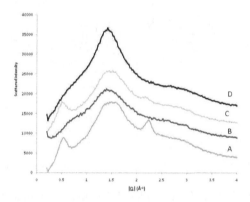

Fig. 9. Wide-angle X-ray scattering curves obtained at room temperature for samples of gelatin (a) cast at 20°C; (b) prepared by electrospinning; (c) sample b held in water vapour for 6 hours; (d) cast at 60°C.

5. Controlling Internal Structure

One of the principle attractions for the use of electrospinning for the preparation of scaffolds is the wide range of materials that can be exploited and the potential for including other processing steps in the scaffold preparation. The properties of solid polymer objects are highly dependent on the structure and morphology present in the material. In turn this is highly dependent on the process conditions. We will now explore some examples of how the structure in the final fibres may be controlled.

5.1 Fibre and Crystal Alignment

A number of reports have highlighted the potential of a rotating collector (Figure 10) to align fibres along a common. Such an approach could be used to develop a higher level of preferred orientation of the semi-crystalline within the nanoscale fibres and hence enhance the properties. Poly(ε-caprolactone) is attractive to scaffold applications as a biomaterial because it is non-toxic. The material is highly elastic compared to other non toxic polyesters and has interesting mechanical properties combining both crystallisation and rubbery behaviour, compatible with, for example, vascular applications. The tensile strengths of electrospun fibres tend to be rather low compared to thicker fibres produced using conventional drawing techniques. Clearly, the control of structure and the semi-crystalline morphology of the fibre material are critical to the process of improving the mechanical properties.

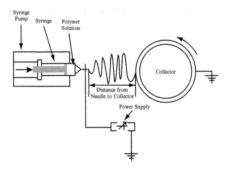

Fig. 10. A schematic of an electrospinning arrangement with a rotating collector

Edwards et al. [22] have explored the use of a rotating collector in the preparation of electrospun fibres of poly(ε-caprolactone). Figures 11a shows a scanning electron microscope image of fibres collected with a rotating electrode with a tangential velocity of 4.3m/s. It is clear that the sample exhibits a very high level of common alignment of the fibres parallel to the tangential velocity vector in contrast to the random matt in Figure 2. We have measured the orientation distributions, $f(\beta)$, for fibres prepared with differing tangential velocities and used these data to calculate the first three orientation parameters, $\langle P_2 \rangle_f$, $\langle P_4 \rangle_f$, $\langle P_6 \rangle_f$ which describe the orientation distribution functions $f(\beta)$ for each set of fibres [29]. The results are plotted in Figure11b as a function of the velocity of the rotating collector surface. For a parallel alignment of the fibres with respect to the rotation direction, $\langle P_2 \rangle_f$, $\langle P_4 \rangle_f$, $\langle P_6 \rangle_f$ will be equal to 1 and for a random orientation these parameters will be equal to 0. The recorded orientation parameters increase with increasing velocity with a peak value at a velocity ~ 2.5 m/s. The values of all three orientation parameters increase towards 1, reflecting a narrowing of the orientation distribution. We have superimposed a plot of the mean diameter of the fibres on Figure 11b. This shows that at the point of the maximum of the fibre orientation at a velocity of ~ 2.5m/s, the fibre diameter has already reduced from the value recorded for a static collector. In other words, the fibre being wound on to the collector is both under tension and has been drawn down. This confirms the expected behaviour that the fibre winding process is affected by the tension and it is the tension which allows a near parallel alignment of the fibres.

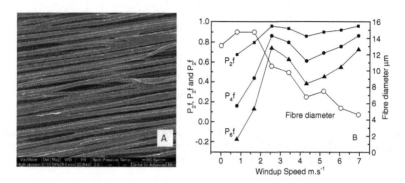

Fig. 11. (a) Scanning electron micrograph of electrospun poly(e-caprolactone) fibres collected on a rotating collector. The direction of rotation is horizontal. (b) Plots of the orientation parameters, $<P_2>$, $<P_4>$, $<P_6>$ describing the angular distribution of electrospun fibres with respect to the velocity direction on the collector [22].

Figure 12 shows the variation of the mean diameter of the electrospun fibres with the tangential velocity of the collector drum. The bars in Figure 12 represent the width of the fibre diameter distribution at half height (FWHM). If we take the density of the fibres to be constant and independent of the rotation speed used in their preparation, we can utilise the fibre diameter data to estimate an effective extension ratio, λ, for the fibres as compared to those prepared with a static collector.

The estimate of the extension ratio, λ, for each sample is also plotted in Figure 12. The plot shows two phases of behaviour. At very low collector speeds there is a little or no variation in the fibre diameter whilst for speeds > 2m/s there is a continual increase in the effective fibre extension ratio with rotator speed, reaching an extension ratio of ~ 10 at a collector speed of 7m/s.

Fibres taken from the rotating collector were examined using small-angle X-ray scattering procedures and Figure 13 shows the SAXS patterns for fibres prepared at four different collector speeds. The SAXS patterns exhibit an increase in preferred orientation with increasing collector speed. The SAXS pattern is typical for a semi-crystalline polymer and shows intense scattering in the form of peaks above and below the zero-angle position which arise from the thin lamellar crystals. The level of preferred orientation can qualitatively be judged by the width of the azimuthal arcing of the peaks. On this basis, the level of preferred crystal orientation is greatest in the sample prepared with a collector speed of 4.3m/s. We have utilised the intensity azimuthal variations to calculate the level of preferred orientation of the crystals using methodologies described elsewhere [22, 29].

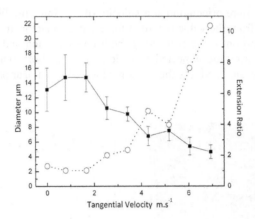

Fig. 12. The mean fibre diameter for poly(ε-caprolcatone) electrospun fibres plotted as a function of the tangential velocity of the collector. The open circles show the extension ratio for each sample calculated assuming a constant density of material [22].

The results are plotted (Figure 14) against the extension ratio as evaluated as described previously. At low extension ratios there are modest level of preferred orientation but the extent of orientation increases with increasing values of the extension ratio and it reaches a maximum for an extension ratio ~ 5. The level of preferred crystal orientation has been deconvoluted from the fibre axis orientation distribution f(β) [22] and thus this can be related to the deformation of an average fibre. In Figure 14 we have compared the measured preferred orientation of the crystals in a fibre with those predicted by the pseudo-affine model for deformation [22]. There is broad similarity in the rate of development of orientation with extension ratio up to ~ 5. Above an extension ratio of 5 the orientation levels observed drop and this attributed to fibre breakage due to the high draw ratio.

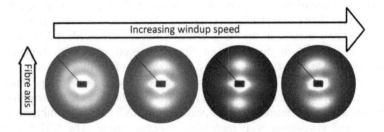

Fig. 13. Small-angle x-ray scattering patterns recorded for electrospun fibres of poly(ε-caprolactone) obtained using different collector speed [22].

We have shown how the three critical parameters of fibre diameter, fibre alignment and alignment of crystals vary with the velocity of the collector. We will consider the variation in this behaviour with increasing collector rotation speed in a number of stages. The first stage corresponds to tangential velocities of less than 2m/s and in this region the 'windup' speed is less than the jet velocity. We deduce this from the observation that there is no draw down of the electrospun fibres (Figure 12).

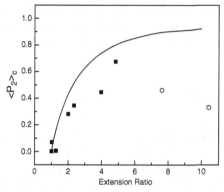

Fig. 14. The symbols show the level of preferred crystal orientation measured with respect to the fibre axis for a series of samples of poly(e-caprolacone) prepared at different collector speeds plotted as a function of the calculated extension ratio. The solid line is the prediction for the level of preferred orientation based on the pseudo-affine model [22].

Fibres which are collected by the rotator will lie at a variety of angles. For tangential velocities greater than 2m/s the fibre diameter reduces with increasing collector speed and hence this places each fibre under a mechanical tension. This favours the alignment of fibres on to the rotating cylinder such that they lie parallel to the rotation direction a behaviour widely observed and modelled in connection with the technology of filament winding. We find that the peak in the orientation distribution of the fibres on the collector occurs at a tangential velocity within this draw down stage, namely 2.5m/s. We have used SAXS to probe the internal structure of the PCL fibres. It is clear that the crystal orientation reaches a maximum at an extension ratio of 5 or a collector velocity of ~ 5m/s and then drops away as the collector rotation speed increases further. We attribute this to breakage of the fibres as the mechanical tension rises with increasing collector speed. PCL is a semi-crystalline polymer and it is well established that the environment present prior to and during the early stages of crystallisation is crucial to templating or directing the crystallisation process. The formation of crystals during electrospinning of PCL solutions will involve the loss of solvent as well as the crystallisation of the polymer itself. If the conformations of some polymer chains are extended during the electrospinning process these may act as row nuclei for the subsequent crystallisation of the remaining polymer, typically the row nuclei are more or less perfectly aligned with respect to the external flow or stress field. If crystals form at an early point of the fibre formation, subsequent extension of the fibre may lead to rotation of the crystals and the development of a significant level of preferred orientation. It is the latter, which is supported by the similarity to the pseudo-affine deformation model in which the crystals rotate affinely with respect to the surrounding deforming matrix but do not change shape.

The generation of an aligned array of electrospun fibres is both critical to some end-use applications and enables the study of those properties in terms of anisotropic properties and structure. This work shows that the velocity of the collector needs careful selection to ensure that the electrospun fibres obtained are representative of the electrospinning rather than any post processing. This is critical in the next section where we study polymer conformations of electrospun fibres using small-angle neutron scattering techniques.

5.2 Chain Alignment

Mohan et al. [23, 24] and Mitchell et al. [30] have used small-angle neutron scattering techniques in conjunction with blends of h-polystyrene and d-polystyrene to probe the shapes and dimensions of polymer chains in electrospun fibres. Such an approach yields amongst other information the radius of gyration of the polymer chain measured both parallel and perpendicular to the fibre axis. Figure 15 shows the radius of gyration obtained for atactic polystyrene electrospun random matts prepared under different conditions plotted as a function of the mean fibre diameter. There is no systematic variation in the radius gyration as the fibre diameter changes. This is much to be expected as the radius of gyration is very much smaller than the fibre diameter. The results cluster around the bulk equilibrium value and suggest the chains are generally in a relaxed state.

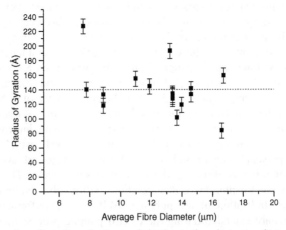

Fig. 15. The radius of gyration of polymer chains in atactic polystyrene electrospun fibres prepared under different conditions plotted as a function of the mean diameter of the fibres. The dashed line is the value obtained from a bulk polystyrene sample. [23, 24]

Mohan et al. also prepared fibres using a rotation collector where the tangential velocity was set equal to the jet velocity. From the aligned arrays of fibres, they were able to measure the ratio of the radius of gyration parallel and perpendicular to the fibre axis, essentially a measure of anisotropy of the chain shape. We might expect that, given the considerable reduction that takes place in diameter from the syringe needle to the final fibre, the level of anisotropy will be very marked. In fact, the ratio plotted as a function of the needle to collector distance (Figure 16) shows remarkably low levels showing that the chains are distorted very little from their quiescent isotropic state. As the electrode separation distance increases, for a fixed applied voltage, the strength of the electric field falls meaning the flight time will be longer.

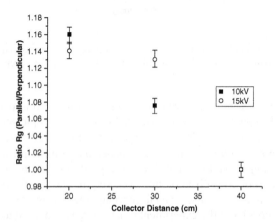

Fig. 16. The ratio of the radius of gyration of polymer chains measured parallel and perpendicular to the fibre axis for atactic polystyrene electrospun fibres prepared with different collector distances for the two applied voltages shown.

Figure 17 shows the ratio of the radius of gyration parallel and perpendicular to the fibre axis plotted as a function of the electric field strength for a set of electrospun fibres. As the electric field strength increases, there is marked increase in the ratio indicating a higher level of anisotropy in the chain trajectories. The level of anisotropy appears to reach a plateau at a field strength of ~ 0.5 kV/cm. Initial analysis suggests that fibres prepared at higher field strengths were subjected to greater deformation.

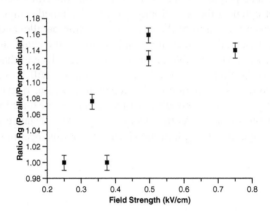

Fig. 17. The ratio of the radius of gyration of polymer chains measured parallel and perpendicular to the fibre axis for atactic polystyrene electrospun fibres prepared with different electric field strengths.

The small-angle neutron scattering studies show for this particular polymer system, that the chains are slightly extended from the isotropic state. This suggests that the deformation of the electrospinning jet for this system takes place when the chains are able to relax, ie there is considerable solvent present. The poly(ϵ-caprolactone) fibres exhibit a high level of crystal orientation and we know from the many studies of crystallisation that even very modest levels of chain extension can lead to nighly oriented crystal morphologies. However, the comparison between the model and the experimental data suggests that the alignment process takes place after the crystals have formed, presumably where there is less solvent.

5.3 Post Processing

The richness of electrospinning partly comes from the possibility of adding active ingredients to the electrospinning solution, such as nutrients for tissue growth, or processing the fibres after spinning to enhance or provide new properties. For example the inclusion of agents to facilitate cross-linking post spinning has been employed by a number of researchers to enhance the mechanical strength of the scaffold. Here, as an example, we describe a process for including reagents in the electrospinning solution which enable the coating of fibres with silver for anti-microbial functionality.

The basis for this methodology is the reaction between ammonical silver nitrate and an aldehyde commonly known as Tollen's test [31]. Sen et al. explored a number of approaches and found the most effective was to incorporate a copolymer based on polystyrene and poly(vinylbenzaldehyde) in to the electrospinning solution. These copolymers were prepared with compositions ranging from 10-50% aldehyde (w/w). They prepared electrospun fibres with diameters of ~2 μm. These fibres were then immersed in Tollen's reagent to produce the silver coating on the fibres. Figure 19 shows scanning electron micrographs of fibres produced. The quality of the silver coating was gauged using energy dispersive x-ray spectroscopy. This was found to be highly dependent on the preparation conditions and the authors report that a high aldehyde content copolymer was required to achieve an even coating.

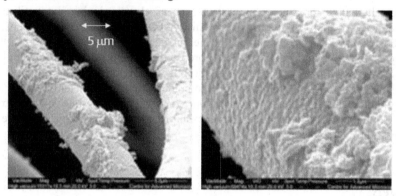

Fig. 18. Electrospun fibres coated with an anti-microbial silver layer using post processing techniques [31]

6 Manufacturing Scale

The vast majority of electrospinning studies and especially that related to the preparation of scaffolds for tissue engineering has been conducted on small-scale laboratory based equipment. There are a variety of approaches which use multiple needles and needle free configurations [32]. A interesting approach is to induce jets through instabilities in the liquid surface through the application of very high electric fields [33, 34]. In any systems with multiple jets there will be complex interactions between the jets and this appears to be an area which would greatly profit from a computational modelling approach in order to explore all of the possible configurations.

7 Computational Modelling and Electrospinning

The previous sections have highlighted the multidisciplinary nature inherent to the use of electrospinning in the fabrication of scaffolds. As with many areas of polymer processing the prediction of properties of a scaffold based on the component materials and the operational parameters and using microscopic theories remains a distant possibility. However, there are a number of well defined phases to the electrospinning process which are tractable in terms of numerical simulation.

7.1 Process Prediction

An alternative for a complex system such as electrospinning is to utilise a neutral network approach to provide a predictive framework for certain properties. Sarkara et al. [35] have exploited the neural network approach to predict the diameters of polyethylene-oxide fibres prepared by electrospinning. Of course such an approach does not incorporate the microscopic physics, but Sarka et al. were able to successfully train the model to predict the fibre diameter based on the four operational parameters: concentration, conductivity, flow rate, and electric field strength. Other approaches, for example using fuzzy logic were less successful [36].

7.2 Jets

The development of the electrospinning jet is fundamental to the process of fabricating electrospun fibres. The physics involved has been addressed by a variety of authors, for example Fridrikh et al. [37] and Reneker et al. [38]. Fridrikh et al. [37] predict that there is a limiting diameter for the fluid jet during the flight path. They develop a model for a Newtonian fluid and neglect the effects which arise due to solvent loss from the jet. A key component in the model is the handling of the chaotic whipping action which takes during flight and leads to a substantial thinning of the fluid jet. They show that the terminal jet radius h_t is given by:

$$h_t = \left(\gamma \bar{\varepsilon} \frac{Q^2}{I^2} \frac{2}{\pi(2\ln\chi - 3)} \right)^{1/3} \tag{2}$$

Where γ is the surface tension, $\bar{\varepsilon}$ the dielectric permittivity, Q the flow rate, I the current and χ is a dimensional wavelength (R/h) of the instabilities defined by the radius R and the jet diameter h. Fridrikh et al. tested this model against experiment observations made while electrospinning poly(ε-caprolactone), polyethylene-oxide, polyacrylonitrile fibre systems. There was broad agreement between experiment and theory and the differences were attributed to the simplicity of the model and the different types of charge carriers.

Reneker et al. [3] have successfully modelled the details of the instabilities in the trajectories of the electrospun fibres and the results compare favourably to data recorded using high speed photography. They were able to model the stable jet and the onset point of instabilities and how these varied with changing operational parameters.

Kumar et al. [39] has extended the consideration of the jet to multiple jets where a novel spinning head was designed to deliver an enhanced electrospinning rate. They compared this novel multi-jet system with a multi-needle electrospinning system by modelling the electric fields in each system and they found that multijet system exhibited a lower degree of fibre repulsion providing a higher throughput..

7.3 Fibres

The development of models to describe the fabrication of fibres using electrospinning range from relatively simple scaling law approaches to detailed simulations involving viscoelasticity, phase separation and solvent evaporation.

Wang et al. [40, 41] have developed empirical scaling laws which relate the operational parameters in electrospinning and the molecular parameters to the jet size and the final fibre diameter. They find that the fibre diameter scales with the solution viscosity with an exponent of 0.41.

Thompson et al. [42] carried out a major study using an electrospinning model to explore the effects of molecular and operational parameters on the final electrospun fibre diameters. They find many parameters effect the diameter but that the five of the parameters (volumetric charge density, collector distance, initial jet radius, relaxation time, and viscosity) have the most effect on the radius. Other parameters (including the polymer concentration, solution density, voltage, and solvent vapor pressure) have a much smaller influence on the radius whilst some parameters such as humidity and surface tension only have modest effect. The study illustrates very effectively the complexity of the electrospinning process and the value of models to bring clarity to such systems.

Dayal and Kyu [43] have simulated the dynamics of the electrospinning process is the situation where the development of the fibre morphology is defined by concentration sweeps. These simulations are complex and involve a variety of elements. They describe the balance of Coulombic and viscoelastic forces at the surface of the jet by using an array of beads connected by Maxwell's elements in a cylindrical shell. Liquid-liquid demixing and solvent evaporation builds on the Cahn-Hilliard approach using a Flory-Huggins free energy. The simulations combine these two processes and generate the key elements of the experimental observations including porosity.

7.4 Mechanical Properties

The mechanical properties of scaffolds are a critical component of the portfolio of properties required. On the macroscopic level the scaffold should have mechanical properties to match those of the tissue, while a smaller scale the fabricated scaffold must exhibit sufficient rigidity to support cell adhesion and extracellular matrix deposition. . Stylianopoulos et al. [44] used a multi-scale modelling strategy to predict the mechanical behaviour of electrospun matts prepared from polyurethane meshes. Experimentally, the measured elastic modulus depended on the fibre diameter and the degree of fibre alignment. Model predictions for the tensile loading parallel to the fibre orientation were in good agreement with experimental measurements. However, the predictions were less accurate in the transverse direction.

There are almost no molecular level simulations in the content of electrospun fibres. Buell et al. [45] has performed detail molecular dynamic simulations at an atomistic level of nanoscale electrospun fibres to explore the very short range interactions between fibres. They use the results to highlight the role temperature in defining the interactions between fibres in non-woven matts.

The examples of computational modelling discussed in the preceding sections underlines the valuable contribution that this approach can make in what often appears to be an intensive experimental activity. There is much to be gained in further modelling activity especially in the fibre production area and the modelling of the evolution nanoscale structure.

8 Conclusions

The process of generating scaffolds for tissue engineering applications using electrospinning techniques offers considerable promises with distinctive advantages over other fabrication methods. We have identified the desirability of using water for biological systems and we show how the control of temperature opens several opportunities as does the inclusion of additives to influence conductivity and viscosity. The use of alternative electrode arrangements can lead to a control of the internal molecular organisation which is vital if target properties are to be achieved. It is fair to see electrospinning as an area dominated by experimental work but there has been from the outset major modelling and theoretical contributions. This computational element continues to grow and its further development will help identify new solutions for the application of electrospinning to tissue engineering.

Acknowledgments

The neutron scattering data were obtained at the STFC ISIS Pulsed Neutron Facility at Chilton UK and the microscopy images were obtained at the Centre for Advanced Microscopy at the University of Reading. We thank our colleagues, in particular, Delyth Elliott and Saeed Mohan, for permission to utilise unpublished data in the preparation of this chapter.

References

1. Greiner A, Wendorff JH. (2007) Electrospinning: a fascinating method for the preparation of ultrathin fibers. Angew. Chem. Int. Ed., 46, 5670-5703
2. Taylor GI. (1964) Disintegration of Water Drops in an Electric Field Proc. R. Soc. Lond. A 280, 383-397
3. Reneker DH, Yarin AL (2000) Bending instability of electrically charged liquid jets of polymer solutions in electrospinning J Appl Phys 87, 4532-4547
4. Sin YM, Hohman MM et al. (2001) 'Experimental Characterization of Electrospinning the Electrically Forced Jet and Instabilities'Polymer 42, 9955-9967
5. Cooley JF (1902) U.S. Patent 692,631
6. Morton WJ (1902) U.S. Patent 0,705,691
7. Lannutti J, Reneker D et al. (2007) Electrospinning for tissue engineering scaffolds. Materials Science and Engineering: C, 27, 504-509
8. Liao S, Chana CK et al. (2008) Stem cells and biomimetic materials strategies for tissue engineering Materials Science and Engineering: C, 28, 1189-1202
9. Liu C, Xia Z et al. (2007) Design and Development of Three-Dimensional Scaffolds for Tissue Engineering Chemical Engineering Research and Design 85, 1051-1064
10. Lewis JA and Gratson GM (2004) Direct writing in three dimensions Materials Today7, 32-39
11. Tellis BC, Sziveka JA et al. (2008) Trabecular scaffolds created using micro CT guided fused deposition modeling Materials Science and Engineering: C, 28, 171-178
12. Williams JM, Adewunmib A, et al. (2005) Bone tissue engineering using polycaprolactone scaffolds fabricated via selective laser sintering Biomaterials 26, 4817-4827
13. Melchels FPW, Feijena J et al. (2009) A poly(d,l-lactide) resin for the preparation of tissue engineering scaffolds by stereolithography Biomaterials 30, 3801-3809
14. Fierz FC, Beckmann F et al. (2008),The morphology of anisotropic 3D-printed hydroxyapatite scaffolds Biomaterials 29, 3799-3806.
15. Nagai Y, Unsworth LD et al. (2005) Slow release of molecules in self-assembling peptide nanofiber scaffold Journal of Controlled Release 115, 18-25
16. Rowlands AS, Lim SA et al. (2007) Polyurethane/poly(lactic-co-glycolic) acid composite scaffolds fabricated by thermally induced phase separation Biomaterials 28, 2109-2121
17. Reignier J and Huneault MA(2006) Preparation of interconnected poly(ε-caprolactone) porous scaffolds by a combination of polymer and salt particulate leaching Polymer, 47, 4703-4717
18. Almeida HA, Bartolo PJ (2009) Computer simulation and optimisation of tissue engineering scaffolds: mechanical and vascular behaviour Proceedings of the 9th Biennial Conference on Engineering Systems Design and Analysis 2, 173-180
19. Lyon J, Li C et al. (2004) Melt-electrospinning part I: processing parameters and geometric properties Polymer 45, 7597-7603
20. Yang L, Fitié CFC et al. (2008) Mechanical properties of single electrospun collagen type I fibers Biomaterials, 29, 955-962
21. Zeugolis DI, Khewd ST et al. (2008) Electro-spinning of pure collagen nano-fibres – Just an expensive way to make gelatin? Biomaterials 29, 2293-2305
22. Edwards MD, Mitchell GR et al. (2010) Development of orientation during electrospinning of fibres of poly(ε-caprolactone) European Polymer Journal 46, 1175-1183
23. Mohan SD , Davis FJ et al. (2009) Electrospinning Atactic-Polystyrene: A Neutron Scattering Study J. Phys.: Conf. Ser. 183 12019 doi:10.1088/1742-6596/183/1/012019
24. Mohan SD , Davis FJ et al. (2010) Modelling Small Angle Neutron Scattering Data from Electrospun Fibres J. Phys.: Conf. Ser. in press
25. Koombhongse M, Liu W et al. (2001) Flat Polymer Ribbons and Other Shapes by Electrospinning Journal of Polymer Science: Part B: Polymer Physics, 39, 2598-2606

26. Shenoy SL, Bates WD et al. (2005) Role of chain entanglements on fiber formation during electrospinning of polymer solutions: good solvent, non-specific polymer–polymer interaction limit Polymer 46, 3372-3384

27. Kariduraganavar MY, Davis FJ et al. (2010) Using an additive to control the electrospinning of fibres of poly(ε-caprolactone). Polymer International 59 827-835 DOI: 10.1002/pi.2795

28. Elliott DE, Davis FJ, et al. (2009) Structure development in electrospun fibres of gelatin J. Phys.: Conf. Ser. 183 12021 doi:10.1088/1742-6596/183/1/012021

29. Mitchell GR, Saengsuwan S,Bualek-Limcharoen S. (2005) Evaluation of preferred orientation in multi-component polymer systems using x-ray scattering procedures Progress in Colloid and Polymer Science 130, 149-159

30. Mitchell GR, Belal M, et al. (2008) Defining structure in electrospun polymer fibres Advanced Materials Research 55-57 33-36

31. Sen S, Davis FJ et al. (2009) Conducting Nanofibres Produced by Electrospinning J. Phys.: Conf. Ser. 183 12020 doi:10.1088/1742-6596/183/1/012020

32. Zhou F-L, Gong R-H (2009) Needle and needleless electrospinning for nanofibers J Appl Polym Sci 115, 2591-2598

33. Jirsak O, Sanetrnik F et al. (2009) Process and apparatus for producing nanofibers from polymer solution electrostatic spinning Patents EP 1673493, CAN 142:393731

34. Jirsak O, Sysel P et al. (2010) Polyamic Acid Nanofibers Produced by Needleless Electrospinning Journal of NanomaterialsVolume , doi:10.1155/2010/842831

35. Sarkara K, Ghaliab MB , et al. (2009) A neural network model for the numerical prediction of the diameter of electro-spun polyethylene oxide nanofibers Journal of Materials Processing Technology 209, 3156-3165

36. Nateri AS, and Hasanzadeh M, (2009) Using Fuzzy-logic and Neural Network Techniques to Evaluating Polyacrylonitrile Nanofiber Diameter Journal of Computational and Theoretical Nanoscience, 6, 1542-1545

37. Fridrikh SV, Yu JH et al. (2003) Controlling the Fiber Diameter during Electrospinning Phys Rev Lett 90 144502-1 144502-4

38. Reneker DH and Yarin AL, (2008) Electrospinning Jets and Polymer Nanofibers Polymer, 49, 2387-2425 DOI:10.1016/j.polymer.2008.02.002.

39. Kumar A, Wei M et al. (2010) Controlling Fiber Repulsion in Multijet Electrospinning for Higher Throughput Macromolecular Materials and Engineering 295, 691-781

40. Wang C, Hsu C-H et al. (2006) Scaling Laws in Electrospinning of Polystyrene Solutions Macromolecules 39, 7662-7672

41. Wang C, Hsu C-H et al. (2008) Scaling Laws and internal structure for characterising in electrospun poly[®-3-hydroxybutyrate] fibers Polymer 49, 4188-4195

42. Thompson CJ, Chase GG, et al. (2007) Effects of parameters on nanofiber diameter determined from electrospinning model Polymer, 48, 6913-6922

43. Dayal P and Kyu T (2007) Dynamics and morphology development in electrospun fibers driven by concentration sweeps Phys. Fluids 19, 107106 doi:10.1063/1.2800277

44. Stylianopoulos T, Bashur CA et al. (2008) Computational predictions of the tensile properties of electrospun fiber meshes: effect of fiber diameter and fiber orientation Computational predictions of the tensile properties of electrospun fiber meshes: effect of fiber diameter and fiber orientation J Mech Behav Biomed Mater. 1, 326–335.

45. Buell S, Rutledge GC et al (2010) Predicting Polymer Nanofiber Interactions via Molecular SImulations ACS Applied Materials & Interfaces 2, 1164-1172

Biofabrication Strategies for Tissue Engineering

Paulo Jorge Bártolo, Marco Domingos, Tatiana Patrício, Stefania Cometa,
Vladimir Mironov

Abstract The success of Tissue Engineering (TE) strongly relies on the capability
of designing biomimetic scaffolds closely resembling the host tissue environment.
Due to the functional multitude of the native tissues, the considerations are
complex and include chemical, morphological, mechanical and biological factors
and their mutability with time. Nonetheless, to trigger and/or assist the "natural
healing mechanism" of the human body it seems essential to provide an
appropriate biomechanical environment and biomolecular signalling to the cells.
Novel biomanufacturing processes are increasingly being recognized as ideal
techniques to produce 3D biodegradable structures with optimal pore size and
spatial distribution, providing an adequate mechanical support for tissue
regeneration while shaping in-growing tissues. In this chapter, we discuss in detail
the most recent advances in the field of biofabrication, providing and updated
overview of processes and materials employed in the production of tissue
engineering constructs. Bioprinting or "scaffold-less" strategies are also
presented in this work. They are based on the precise deposition of high-density
tissue spheroids or cell aggregates being advantageous alternatives to the current
scaffold-based tissue engineering approach.

Paulo Jorge Bártolo
Centre for Rapid and Sustainable Product Development, Polytechnic Institute of Leiria, Centro
Empresarial da Marinha Grande, Rua de Portugal, 2430-028 Marinha Grande, Portugal
email: pbartolo@ipleiria.pt

Marco Domingos, Tatiana Patrício
Centre for Rapid and Sustainable Product Development, Polytechnic Institute of Leiria, Centro
Empresarial da Marinha Grande, Rua de Portugal, 2430-028 Marinha Grande, Portugal

Stefania Cometa
Department of Chemistry & Industrial Chemistry, University of Pisa, 56126 Pisa, Italy

Vladimir Mironov
Advanced Tissue Biofabrication Center, Department of Regenerative Medicine and Cell
Biology, Medical University of South Carolina Charleston, SC 29425, USA

P.R. Fernandes and P.J. Bártolo (eds.), *Advances on Modeling in Tissue Engineering,*
Computational Methods in Applied Sciences 20, DOI 10.1007/978-94-007-1254-6_8,
© Springer Science+Business Media B.V. 2011

1 Introduction

Recent advances in the areas of chemistry, physics, biology, medicine and engineering, and the interconnection between these fields are opening new and exciting perspectives that may revolutionize the actual health services and improve the life quality of worldwide population.

Tissue engineering (TE) is an emerging multidisciplinary scientific field that aims to restore, maintain or enhance tissue and organ functions. TE research includes some of the following areas:

i) Biomaterials: based on novel polymeric, ceramic or hybrid materials that are intended to direct the growth and differentiation of cells forming functional tissues, by providing both physical and chemical cues.

ii) Methodologies for the organization, proliferation and differentiation of cells: by acquiring the appropriate source of cells such as autologous, allogeneic, xenogeneic, stem or genetically engineered cells, and immunological manipulation.

iii) Employment of biologically active molecules: such as growth factors or bone morphogenic proteins.

iv) Engineering design aspects: bi-dimensional cell expansion, three-dimensional tissue growth, bioreactors, vascularization, cell and tissue storage and shipping.

v) Biomechanical design aspects: efficacy and safety of engineered tissues, identification of the minimum required properties for engineered tissues mimicking those of native tissues, mechanical signals regulating engineered tissues.

vi) Informatics: gene and protein sequencing, gene expression analysis, protein expression and interaction analysis, quantitative cellular image analysis, quantitative tissue analysis, digital tissue manufacturing, automated quality assurance systems, data mining tools, and clinical informatics interfaces.

Classical therapeutic strategies in TE field may involve: 1) direct cellular implantation, where cells derived from an endogenous source in the patient or a donor are injected into the damaged tissue; 2) delivery of bioactive factors into the damaged area and 3) *in vitro* combination of cells with a biodegradable scaffold in a dynamic culture system (bioreactor), under conditions designed to engineer a functional tissue for implantation. Once a desired developmental stage is achieved (measured by critical functional properties), the tissue construct can be implanted into the host tissue environment, where further maturation and integration can be obtained.

Several attempts have been made to fabricate artificial extracellular matrices capable of guiding tissue regeneration through time. Conventional techniques have been widely applied in the fabrication of three-dimensional (3D) scaffolds and include solvent casting [1], freeze drying [2-4], phase separation [5], gas foaming [6-7], melt molding [8], fiber bonding [9], particle-leaching techniques [10] and electrospinning [11-12]. Despite being possible to produce porous polymer architectures, with tuneable micro- and macro-scale features, these techniques present several drawbacks, namely the lack of control over pore size, pore interconnectivity, porosity and spatial distribution of the pores. Moreover, the incorporation of cells and proteins within the scaffolds becomes clearly limited by the use toxic organic solvents which are not compatible with most of the biological molecules.

Recently, a huge progress in the biomedical field has been achieved with the introduction of additive fabrication techniques, such as stereolithography [13-14], selective laser sintering [15-16], extrusion-based systems [17-21] and three-dimensional printing [22-23]. These micro-scale technologies, often inheriting the principles of the semiconductor and microelectronics industry, are emerging as a potentially powerful tool for controlling cell microenvironment and generating functional tissue constructs. Besides the high reproducibility and elevated capacity to quickly produce very complex 3D shapes, these techniques enable the fabrication of scaffolds with good control over pore size and spatial distribution increasing the vascularisation and mass transport of oxygen and nutrients throughout the scaffold. Additionally, additive fabrication techniques enable the incorporation of living cells and biologically active molecules during the scaffold fabrication process.

In its traditional meaning, scaffold is intended as a "logistic template", by providing the cells with specific topological features (from nano- to micro- and macro-scale), appropriate biomechanical environment, surface ligands and the ability to release cytokines. As the cells deposit their own extracellular matrix (ECM), the biomaterial scaffold is expected to fully degrade and a tissue-like structure to form and progressively integrate within the surrounding host tissue upon implantation.

In parallel with the abovementioned approaches, some authors have been suggesting ''bioprinting'' as a novel methodology for the construction of 3D biological structures through the computer-aided deposition of cells, cell aggregates and biomaterials [24-25]. For this purpose, commercially available inkjet printers have been successfully redesigned [26] or new ones have been specifically engineered [27-28] to guide biological assembly following a CAD template. These self-assembly and "scaffold-less" approaches are emerging in TE, demonstrating that fully biological tissues can be engineered with specific compositions and shapes, by exploiting cell–cell adhesion and the ability of cultured cells to grow their own ECM, without the employment of synthetic materials, thus reducing the possible inflammatory responses linked to the interaction between biological and non-biological materials.

This chapter provides an updated overview of the most recent advances in the field of biofabrication for TE, making a clear distinction between "scaffold-based" and "scaffold-less" strategies. A critical and extensive description of the currently available fabrication techniques employed in the fabrication of 3D biological substitutes is presented. Bioprinting or "scaffold-less" techniques, based on the biological paradigm, in which the material is assembled bottom up using genetic information to code for an amino acid sequence that contains the information for protein folding and further hierarchical assembly is also reported.

2 Top Down or Scaffold-Based Strategies

2.1 The Scaffolds

The most promising approach in tissue engineering involves the seeding of porous, biocompatible/biodegradable three-dimensional (3D) scaffolds, with donor cells to promote tissue regeneration. Most of the human cell types are anchorage dependent, and therefore scaffolds play a major role in this process, both *in vitro* as well as *in vivo* representing the initial biomechanical support for cell attachment, differentiation and proliferation. The capability to deliver and retain cells and growth factors, enhanced diffusion of cell nutrients, oxygen and vascularisation are critical aspects for a successful organized tissue regeneration process [29-35].

Due to the functional multitude of the tissues it becomes very complex, if not impossible, to establish what defines the ideal scaffold, even for a single tissue type [36]. The considerations are complex and include chemical, morphological, mechanical and biological factors and their mutability with time. Nonetheless, there is a general agreement regarding some of the major requirements [29-32, 34-35]:

i) Biocompatibility – Both raw and processed material should interact positively with the host environment not eliciting adverse host tissue responses. The implantation of an acellular scaffold triggers a sequence of events similar to a foreign body response. The initial acute inflammation response may, in some cases, lead to a chronic inflammatory response that will promote a fibrous capsule development and consequent implant failure [37]. The extent and intensity of the inflammatory response depends on several factors such as defect size, implant biomechanical properties, internal and external architecture, surface properties and porosity [38]. The deviation from the so called normal wound healing process (time and duration) determines the level of biocompatibility of the materials [39]

ii) Biodegradability – Temporary implants (scaffolds) must degrade into non-toxic products with a controlled degradation rate that match the regeneration rate of the native tissue. Ideally, as the scaffold is degraded, cells deposit their own extracellular matrix (ECM) molecules and eventually form 3D structures that closely mimic the native tissue architecture. The *in vivo* degradation process is influenced by different and often conflicting variables, such as the ones related with the material structure (i.e. chemical composition, molecular weight and molecular weight distribution, crystallinity, morphology etc.), its macroscopic features (i.e. implant shape or sizes, porosity etc.) and environmental conditions (i.e. temperature, pH of the medium, presence of enzymes or cells and tissues). Controlling the degradation process and the effects of its degradation products is crucial as the release of acidic degradation products may cause a local drop in pH, which can cause tissue necrosis or inflammation. Several studies have been reported for the degradation of scaffolds of different materials, produced using different techniques [40-43].

iii) Appropriate Porosity, micro and macro structure: Porosity is defined as the percentage of void space in a solid and it is independent of the material [44]. Generally, high level of porosity is required (> 90%) as it increases the surface area, enabling high cell seeding efficiency, migration and proliferation as well as neovascularisation [45]. Besides, adequate porous structure will enhance biomechanical interlocking between the implant and the native tissue at the site of implantation, ensuring greater stability at the interface and avoiding fibrous encapsulation [46]. Nonetheless, porosity itself is not able to guarantee the success of the implant. Adequate pore size and shape (between 100 μm and 500 μm depending upon the tissue) and 100% interconnected network of internal channels are key parameters in terms of cell viability and tissue regeneration. In addition, since the mechanical strength decreases by increasing the porosity, the void volume in implants designed for load-bearing application such as bone, must be balanced in order to accommodate a large number of cells and contemporaneously maintain the required structural strength to avoid the mechanical failure of the implant [47].

iv) Pore size and shape: Pore size plays an important role in terms of cell adhesion/migration, vascularisation and new tissue ingrowth [30, 48-52]. Many researchers have reported optimum pore size ranges according to the different cell lines, tissues and materials. [52-55]. Moreover, the cell colonization of 3D porous matrices is strongly dependent on the ability of cells to bridge pores and spread into the scaffold. If the pores are too large or too small, cells will fail to spread and form networks throughout the scaffold. Therefore it becomes important to define an optimal pore size range for supporting cell and tissue ingrowth [56] Reviewing the literature it seems reasonable that a lower and upper limit, regarding pore size, may be established according to the cell size and maximum void space that cells can span, respectively. These limits greatly vary depending upon cell type and the culture conditions but in general they range between 100μm and 500μm [53].Smaller pores enhance cell adhesion and differentiation *in vitro* while bigger pores promote higher cell adhesion and viability *in vivo*. Despite the great variability observed in the results obtained by different researchers, it seems to be widely accepted that pore size alone is not able to assure an appropriate bone ingrowth and proliferation, which greatly depends also on pore interconnectivity. Moreover, scaffold permeability (provides the means to transport nutrients to, and waste away, from cells) is defined by a combination of five parameters: porosity, pore size and distribution, tortuosity (pore interconnectivity), pore interconnection size and distribution and pore orientation [57-58]. O'Brien et al. [59] studied the effect of pore size on cell adhesion in collagen-gag scaffolds. They found that the fraction of viable cells attached to the scaffold decreased with increasing average pore diameter, and increased linearly with scaffold specific surface area. More recently the same authors conducted a study on the effect of pore size on permeability in collagen scaffolds for tissue engineering. The results of both experimental and mathematical analysis revealed that scaffold permeability increases with increasing pore size and decreases with compressive strain.

v) Mechanical Strength: Implants are required to withstand both *in vitro* manipulation and stresses in the host tissue environment. During *in vitro* culture tissue engineered constructs should maintain their mechanical properties to preserve the required space for cell growth and matrix formation. For *in vivo* applications it is important that constructs mimic as close as possible the mechanical properties of the native tissue providing a temporary mechanical support for tissue regeneration. Initially the scaffold must withstand all stresses and loads in the host tissue environment and then, gradually transfer them to the regenerated tissue. Therefore in the case of polymeric materials and constructs degradation rate represents a key requirement. If the degradation occurs very rapidly physiological stresses may be transferred to the developing tissue prior to sufficient ingrowth and remodelling, resulting in mechanical failure of the repair. On the other hand, slowly degrading materials with high mechanical stiffness can shield the regenerating tissue from external stresses obstructing cell stimulation and tissue regeneration.

vi) Adequate surface finish: an optimal biomechanical coupling between the scaffold and the tissue at the site of implantation, requires an appropriate size and external geometry of the scaffold, and may have a positive effect on the stress distribution at the scaffold/tissue interface [30, 32, 34, 40-41].

vii) Easily manufactured and sterilised: Implants should be rapidly produced with high accuracy and repeatability and easily sterilized by exposure to high temperatures, U.V light or by immersion in a sterilisation agent, remaining unaffected by either of these processes.

viii) Surface topography: Surface roughness may have an important role on the adhesion, proliferation, differentiation and overall cell viability. The scale of the topography (nano to micro scale), shape (e.g., ridges, steps, grooves, pillars and pits) and orientation of the engineered devices may have an effective control on cell behaviour and phenotype expression [60]. In general nano-features (10-300 nm) in the surface of the implants increase cell adhesion and proliferation. Nonetheless these conclusions can't be generalized for all cell phenotypes. Experiments performed with compact carbon nano-tubes revealed a higher number of osteoblasts adhering to the surface of the nano-tubes when compared with fibroblasts, choncrocytes and smooth muscle cells [61]. Despite not being yet clear, it is thought that nano-scale topology modulates the interfacial forces between the cells and the implants. As a consequence, these forces will regulate cytoskeletal formation and organization of cell membrane receptors, trigging complex downstream intracellular molecular signalling [62]. Cell-cell interactions and consequent proliferation/migration may also enhanced/limited by the level of surface roughness. If the roughness level is too high, cells will not be able to establish interconnections and therefore the proliferation/migration will be dramatically compromised. Scaffold surface properties and the interactions established between cells and the ECM are of major relevance in terms of cell signalling and consequent cell adhesion, proliferation and phenotype expression [63-65]. Polymer surface modification is a useful tool to improve the scaffold biofunctionality creating or increasing specific binding sites where bioactive ligands may be immobilized to regulate specific cellular responses [56, 66-67]. Plasma modification processes have been increasingly used in biomedical applications to modify material surface properties, due to their ability to tune in a controlled way, the surface density of different functional groups without altering the substrate mechanical properties [68-69]. Independently of the plasma process used (grafting with non-polymerizable gases like O_2, N_2, NH_3, or polymerization with organic monomers like allyl amine, acrylamide, acrylic acid), substrates displaying surface polar groups (e.g., NH_2, COOH, OH, etc) may alone increase the bioactivity of the substrate and improve cell adhesion and proliferation [70-73]. As previously mentioned, plasma processes can also be quite helpful for the immobilization and deposition of biomolecules such as enzymes, peptides (e.g., RGD: Arginine – Glycine – Aspartic Acid), proteins, polysaccharides and others, onto plasma-modified substrates displaying properly selected binding functional groups [74-48]. The bioactivity of the scaffolds can be increased through the incorporation of growth factors (GF) capable of guiding and controlling cell behavior during tissue regeneration. These signaling molecules are naturally occurring proteins that bind to specific sites and influence cell proliferation and differentiation. In general GF delivery using scaffolds could be achieved either through attachment of the GF onto the scaffolds or physical entrapment within the scaffolds [77]. A wide range of polymeric natural occurring and synthetic materials have been employed to produce drug delivery scaffolds especially for bone and cartilage tissue engineering. The most common and widely applied growth factors in bone and cartilage tissue engineering are vascular endothelial growth factors (VEGF) [78-81], platelet derived growth factors (PDGF) [81-85], nerve growth factor beta (NGF-β) [86-88], transforming growth factor beta (TGF-β) [89-92] different bone morphogenic proteins (BMPs) [93-96] and fibroblast growth factor (FGF) [97-98].

Generally, four classes of biomaterials have been used for engineering tissues: naturally derived polymeric materials (e.g. collagen, gelatine, starch, alginate, and chitosan), acellular tissue matrices (e.g. bladder submucosa and small intestinal submucosa), synthetic polymers (e.g. poly(glycolic acid) (PGA), poly(lactic acid) (PLA), poly(caprolactone) (PCL), poly(lactide-co-glycolide) (PLGA)) and ceramic materials (e.g. hydroxyapatite (HA) and β-tricalcium phosphate (TCP)) [99-104].

Naturally derived polymeric materials and acellular tissue matrices have the potential advantage of biological recognition [104]. Synthetic polymers can be produced reproducibly on a large scale with controlled properties, degradation rate and microstructure.

2.2 Layer-by-Layer Additive Biofabrication

Additive technologies in which physical objects are created directly from computer generated models emerged in the 1980s. The basic concept of additive fabrication is layer laminate manufacturing where three-dimensional (3D) structures are formed by laminating thin layers according to two-dimensional (2D) slice data, obtained from a 3D model created on a CAD/CAM system [105]. Figure 1 provides a general overview of the necessary steps to produce scaffolds for tissue engineering through additive biomanufacturing.

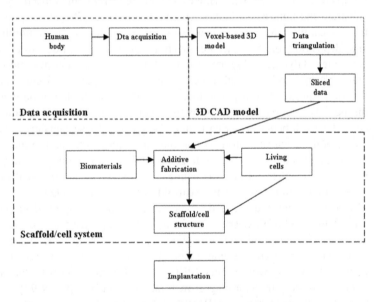

Fig. 1: Steps for additive biofabrication of scaffolds for tissue engineering. Reproduced from [106].

The first step to produce a 3D scaffold through additive biomanufacturing is the generation of the corresponding computer solid model either by the aid of a 3D CAD software or imported from 3D scanners [106]. There are a large number of imaging methods for data acquisition of human or animal body parts, such as X-ray micro computer tomography (μ-CT), Nuclear Magnetic Resonance (MRI), ultrasound (US), single-photon gamma rays (SPECT) and bioluminescence (BLI)

[107-111]. Among these micro-imaging processes, μ-CT is the most commonly used technique to create the scaffold CAD model.

The μ-CT scan produces continuous volumetric data (voxel-based data), which provide the input data for model generation. The model is then tessellated as an STL file, which is currently the standard file for facetted models. Finally, the STL model is mathematically sliced into thin layers (sliced model).

The STL format represents a 3D model by a number of three sided planar facets (triangles), each facet defining part of the external surface of the object [112-114]. There are two kinds of STL files: binary and ASCII files. The difference between these two files is the format of the data definition. A binary file stores the topological information in 32-bit single floating-point format, while the ASCII file stores the information as ASCII strings with keyword strings as indicators [115]. The size of the ASCII STL format is larger than that of binary format but is human readable [112-114]. The structure of an ASCII STL file (Figure 2) starts with the word **solid** followed by the name of the file and ends with the word **endsolid**. Between these two words, the triangles are defined through the specification of the facet normal and the vertices' co-ordinates.

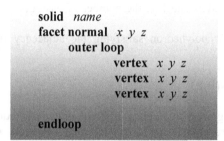

solid *name*
facet normal *x y z*
   outer loop
      vertex *x y z*
      vertex *x y z*
      vertex *x y z*

endloop

Fig. 2: The structure of an ASCII STL file format.

The generation of the STL representation follows two important rules shown in Figure 3 [113, 116]:

- Facet orientation rule: the facets define the surface of the 3D object. The orientation of the facet involves the definition of the vertices of each triangle in a counter-clockwise order;
- Adjacency rule: each triangular facet must share two vertices with each of its adjacent triangles.

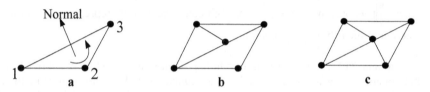

Fig. 3: a) Orientation of a triangular facet in a STL file. b) Violation of vertex-to-vertex rule. c) Correct triangulation.

To assure that the adjacency rule is obeyed it is necessary to verify the following consistency rules [113]:

- The number of triangles must be even;
- The number of edges must be a multiple of three;
- The number of triangles must be equal to two thirds of the number of edges;
- The number of vertices is given by

$$V = 0.5 \times T + 2 \tag{1}$$

where V is the number of vertices and T is the number of triangles.

To assure that the set of STL triangles comprises a closed volume, it is necessary to verify the Euler's rule for solids [117]. According to this rule, the definition of a proper solid implies that [113]:

$$T - E + V - H = 2 \times (B-P) \tag{2}$$

with E being the number of edges, H the number of face holes, B the number of separate, disjoint bodies and P is the number of passages, i.e., holes through the entire body.

For a triangular polyhedron satisfying the adjacency rule, equation (2) is reduced to:

$$-1/2 \ T + V = 2 \times (B-P) \tag{3}$$

For functional graded scaffolds a different strategy is required. In this it must be used a heterogeneous solid model representation describing material composition, grading distribution and geometrical information [118-120]. The space to model functional graded structures is defined as follows:

$$\Omega = G^3 \times M^n \tag{4}$$

where G^3 is the 3D geometry space and M^n is the n-dimensional material space. The material space ξ can be defined as follows:

$$\xi = \left\{ \underline{v} \in M^n, \|\underline{v}\| \equiv \sum_{i=1}^{n} v_i = 1 \, and \, v_i \geq 0 \right\} \tag{5}$$

where $\|.\|$ denotes the L_i-norm and v_i (i-th component of $\underline{v}$) represents the volume fraction of the i-th primary material.

Cai and Xi [121] proposed a morphology-controllable modelling approach for heterogeneous scaffolds based on the shape functions in the finite element method. In this approach the Boolean difference between the contour model of the solid entity and the pore model generates the porous scaffold model (Figures 4 and 5).

Fig. 4: Frustum scaffold modelling. a) the frustum model, b) the whole pore model, c) the resulting frustum scaffold model. Reproduced from [121]

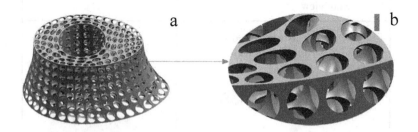

Fig. 5: a) Porous scaffold, generated through the morphology-controllable modelling approach, containing various irregular pores, b) close-up view of a small part of the scaffold. Reproduced from [121]

The main advantages of biomanufacturing techniques are both the capacity to rapidly produce very complex 3D models in a layer-by-layer fashion and the ability to use various raw materials. When combined with clinical imaging data, these fabrication techniques can be used to produce constructs that are customised to the shape of the defect or injury. Some processes operate at room temperature, thus allowing for cell encapsulation and biomolecule incorporation without significantly affecting viability. In the tissue engineering field, biomanufacturing additive processes have been used to produce scaffolds with customised external shape and predefined internal morphology, allowing good control of pore size and pore distribution [103]. These techniques include stereolithographic processes, laser sintering, extrusion and three dimensional printing.

Additive technologies enable the direct and indirect fabrication of scaffolds. The later strategy employs a reverse-lost-mold method in which the scaffold material is cast as illustrated in Figure 6 [122].

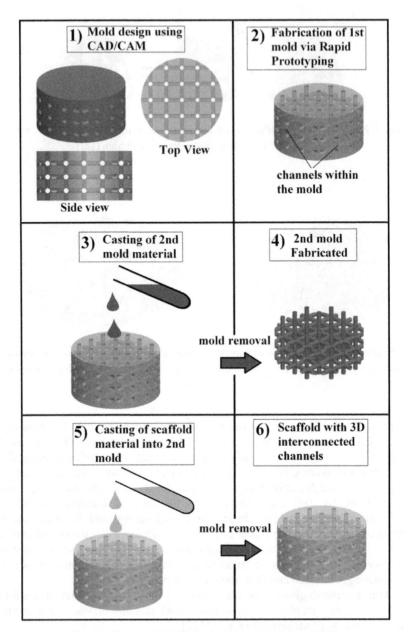

Fig. 6. Indirect scaffold fabrication. Reproduced from [122].

2.2.1 Stereolithographic processes

Stereolithographic processes produce three-dimensional solid objects in a multi-layer procedure through the selective photo-initiated cure reaction of a polymer [105, 123-124]. These processes usually employ two distinct methods of irradiation. The first method is the mask-based method in which an image is transferred to a liquid polymer by irradiating through a patterned mask. The irradiated part of the liquid polymer is then solidified. In the second method, a direct writing process using a focused UV beam produces polymer structures.

The direct or laser writing approach consists of a vat containing a photosensitive polymer, a moveable platform on which the model is built, a laser to irradiate and cure the polymer and a dynamic mirror system to direct the laser beam over the polymer surface "writing" each layer. After drawing a layer, the platform dips into the polymer vat, leaving a thin film from which the next layer will be formed. Mask-based writing systems build models by shining a flood lamp through a mask, which lets light pass through it. These systems generally require the generation of a lot of masks with precise mask alignments. One solution for this problem is the use of a liquid crystal display (LCD) or a digital processing projection system as a flexible mask (Figure 7).

Microstereolithography is a technology that is based on the same manufacturing principle as stereolithography. The first developments of the microstereolithography technique have started in 1993 and different research teams have devised strategies for improving both the vertical and the lateral resolution of the stereolithography process, which resulted in the fabrication of a variety of machines that can be classified into three categories [125-126]:

- **Scanning microstereolithography** machines are based on a vector by vector tracing of every layer of the object with a light beam.
- **Integral mirostereolithography** processes are based on the projection of the image of the layer to be built on the surface of the resin with a high resolution.
- **Two-photon microstereolithography** processes allow the polymerization of the layers composing the object directly inside the reactive medium (and no longer on its surface).

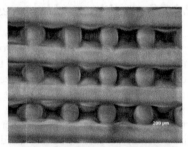

Fig. 7: Mask-based writing system and polyHEMA constructs produced at the Center for Rapid and Sustainable Product Development, Polytechinc Institute of Leiria (Portugal).

Levy et al. [127] used a direct irradiation stereolithographic process to produce HA ceramic scaffolds for orbital floor prosthesis. A suspension of fine HA powder into a UV-photocurable resin was formulated and used as building material. The photo-cured resin acts as a binder to hold the HA particles together. The resin is then burnt out and the HA powder assembly sintered for consolidation. A similar approach was used by Griffith and Halloran [128] that produced ceramic scaffolds using suspensions of alumina, silicon nitride and silica particles with a photo-curable resin. The binder was removed by pyrolysis and the ceramic structures sintered.

Stereolithography is commonly used to produce a negative replica that is filled typically with ceramic slurries and burnt away during sintering. Chu et al. [129] developed a lost-mould technique to produce implants with designed channels and connection pattern. Stereolithography was used to create epoxy moulds designed from negative image of implants. A highly loaded HA-acrylate suspension was cast into the mould. The mould and the acrylic binder were removed by pyrolysis and the HA green scaffold submitted to a sintering process. The finest channel size achieved was about 366 μm and the range of implant porosity between 26% and 52%.

The direct fabrication of biopolymeric scaffolds has also been reported. In another study, Cooke et al. [130] used a biodegradable resin mixture of diethyl fumarate, poly(propylene fumarate) and bisacylphosphine oxide as photoinitiator to produce scaffolds for bone ingrowth. Lan et al. [131] produced poly(propylene fumarate) (PPF) scaffolds with highly interconnected porous structure and porosity of 65%. The scaffolds were coated by applying accelerated biomimetic apatite and arginini-glycine-aspartic acid peptide coating to promote cell behaviour. The coated scaffolds were seeded with MC3T3-E1 pre-osteoblasts and their biologic properties were evaluated using an MTS assay and histological staining. Melchels et al. [132] used a resin based on poly(D,L-lactide) macro-monomers and non-reactive diluent to produce porous scaffolds with gyroid architecture. It was also possible to observe that pre-osteoblasts readily adhered and proliferated well on these scaffolds.

Mask-based writing system can be used to pattern hydrogel structures with high resolution. Liu and Bhatia [133] reported a method, where multiple steps of micropatterned photopolymerisation processes can be coupled to produce 3D cell matrix structures with micro-scale resolution (Figure 8).

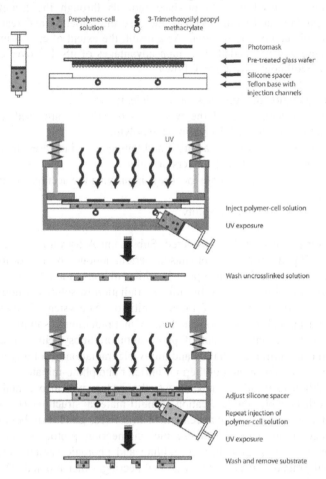

Fig. 8: Process for formation of hydrogel microstructures containing living cells [106].

To produce multimaterial functionally graded scaffolds, researchers from the Centre for Rapid and Sustainable Product Development of the Polytechnic Institute of Leiria (Portugal) are developing a new stereolithographic fabrication process named micro stereo-thermal-lithographic process (µSTLG) [134]. This process uses ultraviolet radiation and thermal energy (produced by IR radiation) to initiate the polymerization reaction in a medium containing both photo- and thermal-initiators (Figure 9). The concentrations of both initiators are carefully selected and the reaction only starts when there is a particular combination of UV radiation and thermal energy [124]. This way, the amount of each initiator must be low to inhibit the start of the polymerisation by only one of these two effects. However, at the point where the two effects intersect each other, the amount of radicals generated is sufficiently high to initiate the polymerisation process. Temperature is used to both produce radicals through the fragmentation of thermal-initiators and simultaneously increase the initiation and reaction rate of the photo-initiated curing reaction. As a result, the extent of cure is increased and no post-cure will be needed. The main advantages of µSTLG over conventional (micro) stereolithography are as follows:

- the generation of radicals is more efficient;
- small concentrations of the two types of initiator are used, enabling the radiation to penetrate deeper into the polymer;
- the combination of UV radiation and temperature increases the reaction rate and hence the fractional conversion values;
- the curing reaction is more localised, improving the accuracy of the produced models;
- the system has more tunability.

Four subsystems can be considered. Subsystem A uses ultraviolet radiation to solidify a liquid resin that contains a certain amount of photo-initiator. This subsystem corresponds to an approach similar to conventional µSL. Subsystem B uses thermal energy produced by infrared radiation to solidify a liquid resin that contains a certain amount of thermal-initiator. Subsystem C uses both heat produced using infrared radiation and ultraviolet radiation to solidify a liquid resin that contains a certain amount of photo-initiator. Subsystem D uses both heat produced using infrared radiation and ultraviolet radiation to solidify a liquid resin containing a certain amount of thermal-initiator and photo-initiator.

In addition to these key advantages, the system will also contain a rotating multi-vat that enables the fabrication of multi-material structures (Figure 10). This represents an important advancement in this field. µSTLG is being developed to produce multi-material microscopic engineering prototypes through nano-structures for exploitation in waveguiding and photonic crystals, multi-material functional graded scaffolds for tissue engineering, other biomedical components and micro-functional metallic or ceramic parts.

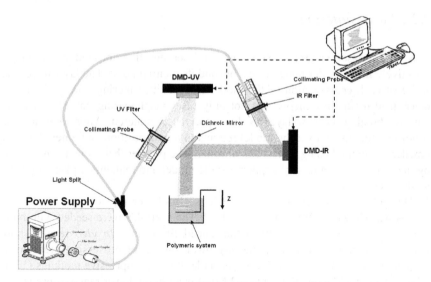

Fig. 9: The micro stereo-thermal-lithographic process: irradiation process.

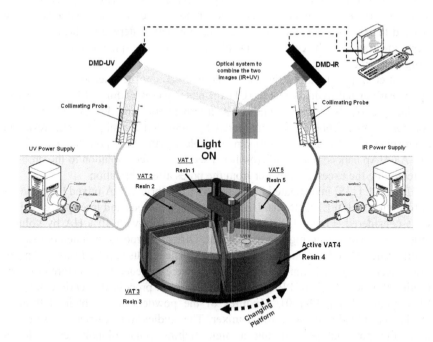

Fig. 10: The micro stereo-thermal-lithographic process: multi-vat system.

2.2.2 Laser sintering

Selective laser sintering (SLS) uses a laser emitting infrared radiation, to selectively heat powder material just beyond its melting point. The laser traces the shape of each cross-section of the model to be built, sintering powder in a thin layer. It also supplies energy that not only fuses neighbouring powder particles, but also bonds each new layer to those previously sintered. After each layer is solidified, the piston over the model retracts to a new position and a new layer of powder is supplied using a mechanical roller. The powder that remains unaffected by the laser acts as a natural support for the model and remains in place until the model is complete.

The potential of SLS to produce PCL scaffolds for replacement of skeletal tissues was shown by Williams et al. [135]. The scaffolds were seeded with bone morphogenetic protein-7 (BMP-7) transduced fibroblasts. *In vivo* results show that these scaffolds enhance tissue in-growth, on top of possessing mechanical properties within the lower range of trabecular bone. Compressive modulus (52 to 67 MPa) and yield strength (2.0 to 3.2 MPa) were in the lower range of properties reported for human trabecular bone.

Lee and Barlow [136] coated calcium phosphate powder with polymer by spray drying slurry of particulate and emulsion binder. The coated powder was then sintered to fabricate calcium phosphate bone implants. Afterwards, these structures were infiltrated with calcium phosphate solution or phosphoric acid-based inorganic cement.

Zhou et al. [137] studied the use of bio-nano-composite microspheres, consisting of carbonated hydroxyapatite (CHAp) nanospheres within a PLLA matrix to produce scaffolds (Figure 11). PLLA microspheres and PLLA/CHAp nanocomposites microspheres were prepared by emulsion techniques. The resultant microspheres had a size of 5-30 μm, suitable for the SLS process. The use of PLLA/CHAp nanocomposite microspheres seems to offer a solution to the problem of removing the excessive powder from the pores after fabrication.

Hao et al. [138] investigated the use of SLS to fabricate HA mixed high density polyethylene (HDPE) scaffolds. Different scanning speeds and laser power values were considered (Figure 12). HA and HDPE powders with 40% HA by volume ratio were mixed using high speed blender. Different process parameters resulted in different sintered morphologies (Figure 13). The results revealed shown that for low power or high scanning speed the layers were in general not sintered or very fragile. Powder blends of PEEK/HA have also been processed experimentally on SLS (Figure 14) [139-140]. Mixtures of the powders were obtained through mechanical blending using a roller-mixer. The studies were carried out to determine the potential of sintering a high melting point biopolymer in lower temperature environment.

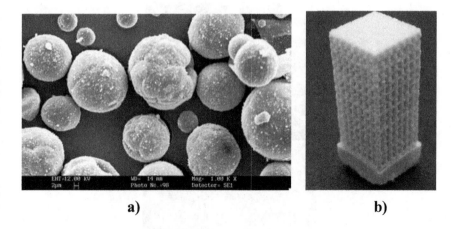

a)   b)

Fig. 11: a) SEM image of PLLA/CHAp nanocomposite microspheres. B) PLLA/CHAp nanocomposite scaffolds [137].

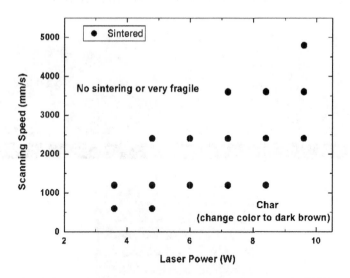

Fig. 12: Processing window for the SLS of HA/HDPE composite [138].

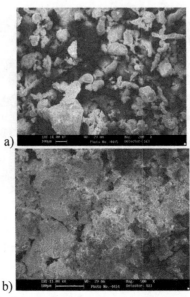

Fig. 13: SEM micrograph of sintered layer's surface with irradiated at 1200 mm/s of scanning speed and different laser power: (a) 3.6W. (b) 7.2W [138].

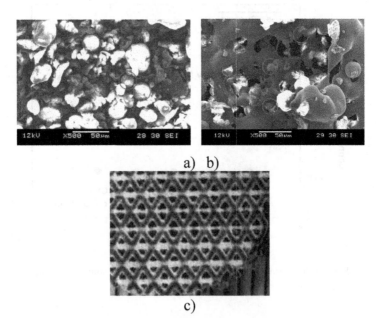

a) b)

c)

Fig. 14: SEM micrographs of PEEK-10% wt HA (a) before and (b) after sintering. (c) Bottom view of PEEK-HA composite scaffold [139-140].

2.2.3 Extrusion-based processes

The extrusion-based technique, commercially known as Fused Deposition Modelling (FDM), was developed by Crump [141]. By this process, thin thermoplastic filaments are melted by heating and guided by a robotic device controlled by a computer, to form the object. The material leaves the extruder in a liquid form and hardens immediately. The previously formed layer, which is the substrate for the next layer, must be maintained at a temperature just below the solidification point of the thermoplastic material to assure good interlayer adhesion.

Extrusion-based processes have been used to successfully produce scaffolds in PCL, PP-TCP, PCL-HA, PCL-TCP with resolution of 250 μm. Some of the major limitations of FDM are due to the use of filament-based materials and the high heat effect on raw material. In order to solve some limitations of the FDM process, such as the requirement of precursor filaments or high processing temperatures, some alternative processes have been proposed.

Hutmacher et al. [142] optimised the FDM processing parameters for the production of PCL scaffolds. *In vitro* cell response of primary fibroblasts and osteoblast cells on PCL scaffolds was investigated. Similar work was conducted by Zein et al. [143] that produced PCL scaffolds with a range of channel size 160-700 μm, filament diameter 260-370 μm, porosity 48-77% and regular honeycomb pores. The compressive stiffness ranged from 4 to 77 MPa, yield strength from 0.4 to 3.6 MPa, and yield strain from 4% to 28%.

Woodfield et al. [144] used a FDM-like technique, called 3D Fibre Deposition, to produce poly(ethylene glycol)-terephthalate-poly(butylenes terephthalate) (PEGT/PBT) block co-polymer scaffolds with a 100% interconnecting pore net-work for engineering of articular cartilage. By varying the co-polymer composition, porosity and pore geometry, scaffolds were produced with a range of mechanical properties close to articular cartilage. The scaffolds seeded with bovine chondroccytes supported a homogeneous cell distribution and subsequent cartilage-like tissue formation.

Drexel University developed a variation of FDM called Precision Extruding Deposition for fabrication of bone tissue scaffolds. In this process, material in pellet or granule form is fed into a chamber where it is liquefied. Pressure from a rotating screw forces the material down a chamber and out through a nozzle tip. This process was used by Wang et al. [145] to directly fabricate PCL scaffolds with controlled pore size of 250 μm and designed structural orientations (0°/90°, 0°/120° or both combined patterns). Proliferation studies were performed using cardiomyoblasts, fibroblasts and smooth muscle cells. The surface hydrophilicity and total surface energy of PCL scaffolds was also increased with plasma treatment [146]. Different plasma treatments duration (1, 2, 3, 5 and 7 minutes) were performed. Results show that the maximum value of total surface energy (Figure 15) and its components (polar and dispersive) occur for 3 minutes of treatment. Additionally, it was observed strength of cell adhesion, which increased

55% on 3 minutes plasma-treated scaffolds compared to untreated and other plasma treatment times [146].

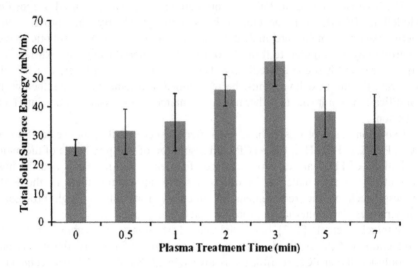

Fig. 15: Plasma treatment of PCL scaffolds: the total energy (mN/m) for various plasma treatment times [146].

Park et al. [147] compared pore interconnectivity and the biological characteristics of PCL scaffolds produced by melt extrusion and salt-leaching. The scaffolds were seeded with chondrocytes. Similarly, Rath et al. [148] used a robotic dispensing system to produce PCL-PEO scaffolds in two lay-out patterns (0°/90° and 0°/45°/90° with porosity values of 60% and 71% respectively) using a 0.5 mm nozzle. The scaffolds were seeded with porcine articular chondrocytes along with two different hydrogels (alginate and alginate/thrombin). From the experimental work it was possible to observe that the 0°/45°/90° pattern showed increased cell numbers and extracellular matrix proteins such as collagen-II. The alginate/thrombin matrix encapsulated more chondrocytes.

In order to eliminate the elevated temperatures required by the extrusion-based processes, Tsinghua University researchers developed a process called Low-temperature Deposition Manufacturing (LDM) to produce scaffolds at a low temperature environment under 0°C [149]. The LDM system comprises a multi-nozzle extrusion process and a thermally induced phase separation process (Figure 16). Scaffolds having a macroporous structure larger than 100 μm in diameter and a microporous structure smaller than 100 μm have been reported. The LDM process was used to produce poly(L-lactide) (PLLA) and TCP composite scaffolds with BMP growth factor. The scaffolds were implanted into rabbit radius and canine radius with large-segmental defects. After 12 weeks it was possible to observe that the rabbit radius defect was successfully repaired and the regenerated bone had properties similar to the healthy bone. For the canine radius similar results were observed after 24 weeks [149].

An alternative process is the pressure assisted microsyringe (PAM) that involves the deposition of polymer dissolved in solvent through a syringe [150]. The thickness of the polymer stream can be varied by changing the syringe pressure, solution viscosity, syringe tip diameter and motor speed. Resolution as low as 10 μm on a 2D structure was achieved.

Robocasting (Figure 17), also known as direct-write assembly, consists on the robotic deposition of highly concentrated colloidal suspensions capable of fully supporting their own weight during assembly due to their viscoelastic properties [151]. This technique have been used to produce β-tricalcium phosphate (β-TCP) scaffolds [151-152].

The Polytechnic Institute of Leiria developed a variation of FDM called BioExtruder [43, 154-156]. It is a highly reproducible and low cost system enabling the controlled definition of pores into the scaffold to modulate mechanical strength and molecular diffusion, as well the fabrication of multi-material scaffolds. It comprises two different deposition systems: one rotational system for multi-material deposition acted by a pneumatic mechanism and another one for a single material deposition that uses a screw to assist the deposition process (Figure 18). The rotational system has four reservoirs, two with temperature control and two without. A large number of nozzle diameters ranging from 0.1 to 1 mm can be used. Three scanning strategies were implemented for the deposition process: contour, raster, and contour and raster. The deposition code, developed in Matlab (The Math-Works, Inc.), is based on the ISO programming language commonly used to control Computer Numerical Control (CNC) machines. Poly(ε-caprolactone) was the material chosen to produce porous scaffolds, made by layers of directionally aligned microfilaments. Chemical, morphological, and *in vitro* biological evaluation performed on the polymeric constructs revealed a high potential of the BioExtruder to produce 3D scaffolds with regular and reproducible macropore architecture, without inducing relevant chemical and biocompatibility alterations of the material [155]. Several control parameters such as temperature, screw rotation velocity, deposition velocity and slice thickness and their direct influence on morphological and mechanical properties of the extruded scaffolds were studied [156]. Fibroblasts were successfully attached to PCL scaffolds produced through the Bioextruder as shown in Figure 19 [155].

The *in vitro* degradation behaviour in both simulated body fluid (SBF) and phosphate buffer solution (PBS) were also investigated. The characterization of the degradation behaviour of the structures was performed at specific times by evaluating changes in the average molecular weight, the weight loss and its thermal properties. Morphological and surface chemical analyses were also performed using a Scanning Electron Microscopy (SEM) and an X-ray Photoelectron Spectroscopy (XPS), respectively [43].

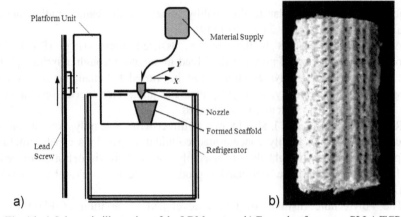

Fig. 16: a) Schematic illustration of the LDM system. b) Example of a porous PLLA/TCP composite scaffold produced by LDM process [153].

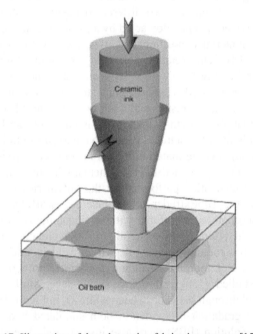

Fig. 17: Illustration of the robocasting fabrication process [151].

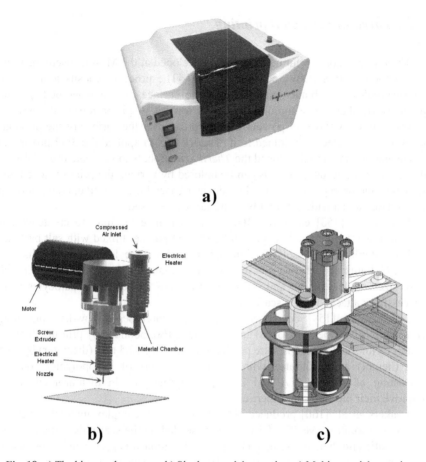

Fig. 18: a) The bioextruder system; b) Single-material extrusion; c) Multi-material extrusion system.

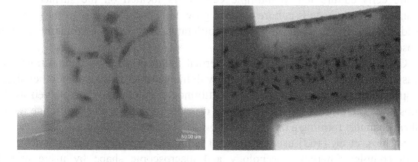

Fig. 19: Bright field micrographs of mouse embryo fibroblasts stained with toluidine blue after six days of culture onto PCL-scaffolds [155].

2.2.4 Three-dimensional printing

Three-dimensional printing (3DP) was developed at the Massachusetts Institute of Technology (USA) by Sachs et al. [157]. The process deposits a stream of microparticles of a binder material over the surface of a powder bed, joining particles together where the object is to be formed. A piston lowers the powder bed so that a new layer of powder can be spread over the surface of the previous layer and then selectively joined to it. Therics (USA) applied the 3DP process to tissue engineering and developed the *TheriForm* process to fabricate drug delivery devices and scaffolds. Scaffolds can be tailored by varying the printing speed, the flow rate and drop position of the liquid water-based binder. 3DP operates at room temperature and a wide range of biomaterials can be used.

Kim et al. [158] employed 3DP with particulate leaching to create porous scaffolds, using polylactide-coglycolide (PLGA) powder mixed with salt particles and a suitable organic solvent. The salt particles were leached using distilled water. Cylindrical scaffolds measuring 8 mm (diameter) by 7 mm (height) with pore sizes of 45-150 μm and 60% porosity were fabricated. Hepatocytes were successfully attached to the scaffolds.

Lam et al. [159] developed a blend of starch-based powder containing cornstarch (50%), dextran (30%) and gelatine (20%), bounded by printing distilled water. Cylindrical scaffolds were produced measuring 12.5 mm (diameter) by 12.5 mm (height) and infiltrated with different amounts of a copolymer solution consisting of 75% L-PLA and 25% polycaprolactone in dichloromethane to improve their mechanical properties.

Leukers et al. [160] produced HA scaffolds with complex internal structures and high resolution. MC3T3-E1 cells were seeded on the scaffolds and cultivated under static and dynamic setups. Dynamic cultivation was performed in perfusion containers. A flow rate of 18 μl/min. Histological evaluation was carried out to characterise the cell ingrowth process. It was observed that the dynamic cultivation method lead to a stronger population compared to the static cultivation method. Static cells culture led to multiple cell layers located on the surface of HA granules. Dynamic cells culture tends to grow in between cavities of the granules. Additionally, it was found that cells proliferated deep into the structure forming close contact to HA granules.

Cui et al. [161] used a modified thermal inkjet printer and demonstrated the feasibility of printing microvasculature with human microvascular endothelial cell suspension in thrombin solutions onto fibrinogen solutions which served as the substrate. The printed cells achieved the capacity to interact and proliferate within fibrin channels forming a tubular lining.

Sachlos et al. [162] used an indirect approach to produce collagen scaffolds with complex internal morphology and macroscopic shape by using a 3DP sacrificial mould. A dispersion of collagen was cast into the mould and frozen. The mould was then dissolved with ethanol and the collagen scaffold was critical point dried with liquid CO_2. Similarly, Lee et al. [163] used indirect 3DP technology to produce scaffolds with large pore sizes with fine features. The ability of these scaffolds to support intestinal ephithelial cell culture was investigated.

3 Bottom Down or Scaffold-Less Strategies

Organ printing is a rapid emerging variant of the biomedical application of rapid prototyping or additive manufacturing to the science of tissue engineering. Organ printing can be defined as a computer-aided robotic-by-layer additive biofabrication of 3D functional human tissue and organ constructs using self-assembling tissue spheroids as building blocks. Organ printing includes three essential steps: (i) pre-processing or design of an "organ blueprint"; (ii) processing or bioprinting of 3D tissue and organ constructs using tissue spheroids as "bioink"; and (iii) post-processing or accelerated tissue maturation. Using an analogy of text printing, it is possible to identify several critically important elements of organ printing technology: "blueprint" or computer aided design of human organ construct in STL file, "bioink" or self-assembling tissue spheroids, "biopaper" or bioprocessible and biomimetic hydrogel, and a "bioprinter" or robotic dispenser. Bioprinting a "built in" perfusable intraorgan branched vascular tree is critically important for maintaining the viability of bioprinting 3D thick human organ constructs. Combined employment of solid, mono-lumenal and multi-lumenal tissue spheroids is sufficient to bioengineer a complete intraorgan branched vascular tree. The feasibility of the bioassembly of linear and branched segments of a vascular tree, using vascular tissue spheroids, has already been demonstrated. However, development of novel technologies for scalable biofabrication of tissue spheroids as well as new methods for accelerated tissue maturation are critical for successful clinical translation and commercialization of organ printing technology. In the short term, this emerging organ printing technology will be instrumental for robotic biofabrication of 3D models of vascularized human tissues and organs for modeling human diseases, drug discovery and toxicology research. In the long term, organ-printing technology can solve one of the most pressing clinical challenges: the shortage of human organs for transplantation.

Tissue engineering scientists are increasingly using robotic methods to develop human tissue constructs. Often researchers will use rapid prototyping to manufacture scaffolds on which to seed human cells, or robotically spray cells and biomaterials to create tissue constructs. Recent developments have focused on using robotic methods to accurately deposit cells and biomaterials. The first development, and most common is the precise placement of cells using ink jet printing, laser deposition or stereolithography methods. The second approach has been to produce cell-hydrogel mixtures using embedding and molding, cell sheeting or electrospinning technology. The third approach to precisely position cell and biomaterials has been to dispense high-density tissue spheroids or cell aggregates using cell spheroids as building blocks. Other methods to position cells and biomaterials involve using robotic means to control cell migration. These techniques include dielectrophoresis and magnetic force driven biofabrication [164]. Each of these methods offers a series of benefits and limitations, from which no clear leader has emerged in the manufacture of human tissue constructs.

Organ printing, or bioprinting technology has developed around the approach to precisely position high-density tissue spheroids or cell aggregates. Bioprinting offers an advantageous alternative to solid scaffold based tissue engineering; (i) organ printing is an automated pathway for the scalable reproduction of mass produced tissue engineered products from standardized modular blocks; (ii) bioprinting allows for the precise placement and three-dimensional positioning of several cell types due to the XYZ coordination of the hardware; (iii) organ printing allows for a high level of cell density, especially when using cell aggregates as the modular building blocks or bioink; (iv) bioprinting has the potential to solve the problem of vascularization through the deposition of a vascular tree within the construct; (v) tissue engineers now believe that tissue and organs are self-organizing systems that normally undergo biological self-assembly and organization without the need for solid scaffolds (Figure 20).

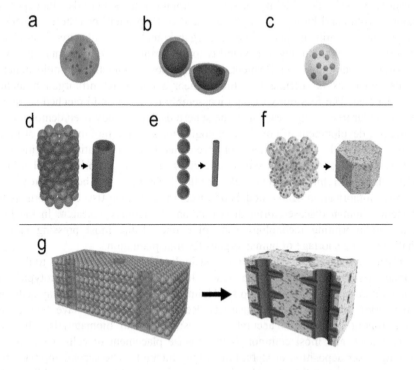

Fig. 20: Organ printing principles and methodologies.

As stated by David Williams in the journal Biomaterials, 'It might seem at first sight that if tissue engineering can be achieved by self-assembly of tissue components without the need for conventional solid materials, biomaterials would have no further role. In line with many earlier statements... I would consider the self-assembled tissue to be an engineering construct, and therefore, a biomaterial in its own right' [165]. Organ printing using self-organizing tissue spheroids may remove the need for scaffolds and biomaterials altogether.

Another potential advantage of the minitissue-based approach is its promise to solve the problem of vascularization. For optimal perfusion, organs such as the kidney need a well-developed branched intraorgan vascular tree. To design an effective perfusion process of thick 3D macrotissue it is important to build a dense network of interconnected roads without highways. Research into this area has shown that microvascular networks can self-assemble from either single cells or from endothelial tissue spheroids (Figure 21). Further, it has been shown that small isolated fragments of the microvascular tree can be reunited by self-assembly in vitro and in vivo [166]. Thus we can argue that building an intraorgan branched vascular tree with 10-12 orders is technologically feasible. This type of organ printing involves printing soft physical 'voxels' that can self align with its neighbors while being capable of self-assembly. Tissue spheroids are considered beneficial for this reason and because they are easy to fabricate at a desirable standard size at large scale.

In order to achieve the biomechanical properties of tissue engineered constructs it is essential to know in advance not only the material properties of adult tissue and organs, but also the development kinetics and the structural determinants of the materials properties at different stages of development. Any bioprinting technology must deal predominantly with the enhancement of synthesis, deposition, assembly, cross linking and intermolecular interactions of the extracellular matrix molecules especially in the case of organs consisting of dense connective tissue. In the absence of solid scaffold, bioprinted tissue constructs must also undergo rapid maturation. The development of an irrigating bioreactor with removable porous microtubes could help to buy time until the tissue matures. Alternatively, the use of sacrificial hydrogels such as hyaluronon (degraded by hyaluronidase) could facilitate the temporary delivery of nutrients and rigidity to the tissue construct [166]. Finally, the recruitment of circulated extracellular matrix molecules such as collagen type 1 and fibronectin could aid the maturation of structural integrity of the tissue construct.

The recent developments in bioprinting research and analysis of the future have reveal that the progress of organ printing in tissue engineering will depend on:

1. How quickly we can switch from using synthetic biomaterials to using living natural tissues as biomaterials and building blocks.
2. How quickly we can switch from bioprinting in vitro to bioprinting in vivo.

3. How quickly we can switch from 'top down' (solid scaffold based) to 'bottom up' (self-assembly-based) or modular approach employing scalable computer-aided, automated robotic biofabrication systems.
4. How quickly we can evolve from analog (continuous) bioprinting to digital (droplet) bioprinting.
5. The effectiveness of adaptation and incorporation of novel technologies developed outside of the traditional tissue engineering research domains:
   i.   Nanotechnology and nanomedicine.
   ii. Continuous and digital microfluidics.
   iii. Synthetic biology and macromoluecular chemistry.
   iv. Stem cell, developmental and regenerative biology.
   v.   Computational biology and computer science.
   vi. Biomaterial science, biomechanics, mechanobiology.
   vii. Robotics & mechanical engineering and so on...

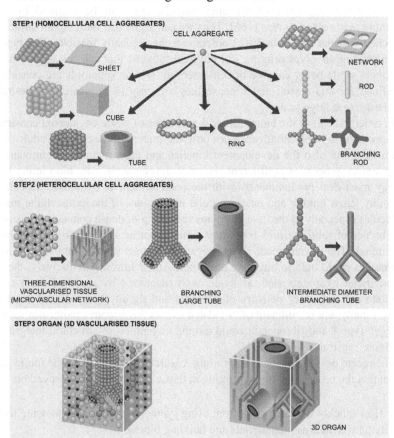

Fig. 21: Microvascular networks self-assemble process.

The long-term success of bioprinting depends on the pull from the potential uses and users of the technology. Some believe that bioprinting can offer the solution to the main limitation of biofuels, by bioprinting algae into biofilms [167]. Some researchers also consider bioprinting for the large scale-production of animal free food, leather and furs to reduce to demand for animals and increase the supply of animal produce. Whilst each of these applications has good merit, the most commonly recognized drivers are within the clinical market. Bioprinted human tissues could offer better predictability for pharmacological and toxicological research that either 2D assays and animal models. Market research prediction estimate that this application will generate over $10billion p.a. within the next 20 years. Further, the clinical market for human organs suitable for transplantation is also growing because of a shortage of donors and a greater number of people needing transplants. Market research estimates that the market for bioprinted organ will surpass $30billion within the next 30 years. Lastly, the most lucrative market, and perhaps the least recognized is the market for *in vivo* bioprinting to replace damaged, injured or lost human tissue. Several groups from around the work have demonstrated the feasibility of *in vivo* bioprinting. Dr Fabrien Guillemot from the University of Bordeaux, France has printed nano-hydroxapatite into a mouse bone defect model [168]. Dr Laurence Bonassar from Cornell University printed cartilage cells directly onto isolated bone [169] and Prof. Anthony Atala from Wake Forest University recent reported the successful testing of ink jet skin printing onto a animal to replace burnt tissue. Estimates predict that this market will generate upwards of $50billion within the next 30 years.

In summary organ printing or biomedical rapid prototyping is the new science driving tissue engineering. Much like text printing, organ printing requires a tissue 'blueprint,' bioink, biopaper and a bioprinter, or robotic dispenser. Tissue engineering scientists are increasingly adopting robotic methods to develop human tissues. The bioprinting technique has gained popularity for its scalability, ability to precisely place several cell types, and potential to solve the problem of vascularization. Nevertheless, further research is needed to reproduce the essential features of normal natural tissues. How closely can tissue engineers capture the most essential structure-function features of normal natural tissues and organs, and how far must they go to imitate these processes is still open for debate. What is clear is that the progress of organ printing will follow specific rules based on the limitations of current knowledge. The technological feasibility and the market drivers suggest that organ printing is achievable, and all that is needed now is time.

References

1. Schmitz JP, Hollinger JO (1988) A preliminary study of the osteogenic potential of a biodegradable alloplasticosteoinductive alloimplant, Clin Orthop Relat Res, 237, 245-551.
2. Whang K, Thomas CH, Healy KE and Nuber G (1995) A novel method to fabricate bioabsorbable scaffolds, Polymer, 36, 837-42.

3. Hsu YY, Gresser JD, Trantolo DJ, Lyons CM, Gangadharam PR and Wise DL (1997) Effect of polymer foam morphology and density on kinetics of *in vitro* controlled release of isoniazid from compressed foam matrices, J Biomed Mater Res, 35, 107-116.

4. Schoof H, Apel J, Heschel I and Rau G (2001) Control of pore structure and size in freeze-dried collagen sponges, J Biomed Mater Res, 58, 352-357.

5. Lo HP and Leong KW (1995) Fabrication of controlled release biodegradable foams by phase separation Tissue Eng, 1, 15-28.

6. Mooney DJ, Baldwin DF, Suh NP, Vacanti JP and Langer R (1996) Novel approach to fabricate porous sponges of poly(D, L-lactic-co-glycolic acid) without the use of organic solvents, Biomaterials, 17, 1417-1422.

7. Nazarov R, Jin HJ and Kaplan DL (2004) Porous 3D scaffolds from regenerated silk fibroin, Biomacromolecules, 5, 718-726.

8. Thompson RC, Yaszemski MJ, Powers JM and Mikos AG (1995) Fabrication of biodegradable polymer scaffolds to engineering trabecular bone, J Biomater Sci-Polym, 7, 23-38.

9. Cima LG, Vacanti JP, Vacanti C, Inger D, Mooney DJ and Langer R (1991) Tissue engineering by cell transplantation using degradable polymer substrates, J Biomech Eng, 113, 143-151.

10. Hofmann S, Hagenmuller H, Koch AM, Muller R, Vunjak-Novakovic G, Kaplan DL, Merkle HP and Meinel L (2007) Control of *in vitro* tissue engineered bone-like structures using human mesenchymal stem cells and porous silk scaffolds, Biomaterials, 28, 1152-1162.

11. Li WJ, Laurencin CT, Caterson EJ, Tuan RS and Ko FK (2002) Electrospun nanofibrous structure: a novel scaffold for tissue engineering, J Biomed Mater Res, 60, 613-621.

12. Ma Z, Kotaki M, Inai R and Ramakrishna S (2005) Potential of nanofiber matrix as tissue-engineering scaffolds, Tissue Eng, 11, 101-109.

13. Fedchenko F (1996) Stereolithography and other RP&M technologies, Edited by PF Jacobs, ASME Press.

14. Melchels FPW, Feijen J and Grijpma DW (2010) A review on stereolithography and its applications in biomedical engineering, Biomaterials, 31 (24), 6121-6130.

15. Hutmacher DW, Cool S (2007) Concepts of scaffold-based tissue engineering - the rationale to use solid free-form fabrication techniques, J Cell Mol Med, 11, 654-669.

16. Yeong W, Chua C, Leong K and Chandrasekaran M (2004) Rapid prototyping in tissue engineering: challenges and potential, Trend Biotechnol, 22, 643-652.

17. Miot S, Woodfield T, Daniels AU, Suetterlin R, Peterschmitt I, Heberer M, van Blitterswijk CA, Riesle J and Martin I (2005) Effects of scaffold composition and architecture on human nasal chondrocyte redifferentiation and cartilaginous matrix deposition, Biomaterials, 26, 2479-2489.

18. Hutmacher DW, Kirsch A and Ackermann KL (2001) A tissue engineered cell-occlusive device for hard tissue regeneration—a preliminary report, Int J Periodontics Restorative Dent, 21, 49-59.

19. Hoque E, San WY, Wei F, Li S, Huang M-H, Vert M, Hutmacher DW (2009) Processing of polycaprolactone and polycaprolactone-based copolymers into 3D scaffolds, and their cellular responses, Tissue Engineering: Part A, 15 (10), 3013-3024.

20. Landers R and Mulhaupt R (2000) Desktop manufacturing of complex objects, prototypes and biomedical scaffolds by means of computer-assisted design combined with computer-guided 3D plotting of polymers and reactive oligomers, Macromol Mater Eng, 282, 17-21.

21. Landers R, Hubner U, Schmelzeisen R and Mülhaupt R (2002) Rapid prototyping of scaffolds derived from thermoreversible hydrogels and tailored for applications in tissue engineering, Biomaterials, 23, 4437-4447.

22. Shen F, Cui YL, Yang LF, Yao KD, Dong XH, Jia WY and Shi HD (2000) A study on the fabrication of porous chitosan/gelatin network scaffold for tissue engineering, Polym Int, 49, 1596.

23. Sherwood JK, Riley SL, Palazzolo R, Brown SC, Monkhouse DC, Coates M, Griffith LG, Landeen LK and Ratcliffe A (2002) A three-dimensional osteochondral composite scaffold for articular cartilage repair, Biomaterials, 23, 4739-4751.

24. Mironov V, Boland T, Trusk T, Forgacs G and Markwald RR (2003) Organ printing: computer-aided jet-based 3D tissue engineering, Trends Biotechnol, 21, 157-161.

25. Ringeisen BR, Othon CM, Barron JA, Young D and Spargo BJ (2006) Jet-based methods to print living cells. Biotechnol J, 1, 930-948.

26. Boland T, Xu T, Damon B, Cui X (2006) Application of inkjet printing to tissue engineering, Biotechnol J,1, 910-917.

27. Nakamura M, Kobayashi A, Takagi F, Watanabe A, Hiruma Y, Ohuchi K, Iwasaki Y, Horie M, Morita I and Takatani S (2006) Biocompatible inkjet printing technique for designed seeding of individual living cells, Tissue Eng, 11, 1658-1666.

28. Saunders RE, Gough JE and Derby B (2008) Delivery of human fibroblast cells by piezoelectric drop-on-demand inkjet printing, Biomaterials, 29, 193-203.

29. Chang R, Sun W (2009) Biofabrication of three-dimensional liver cell-embedded tissue constructs for in vitro drug metabolism model, LAP Lambert Academic Publishing.

30. Bártolo PJ, Almeida HA, Rezende RA, Laoui T and Bidanda B (2008) Advanced processes to fabricate scaffolds for tissue engineering, Virtual Prototyping & Bio-manufacturing in medical applications, Edited by PJ Bártolo and B Bidanda, Springer.

31. Holtorf HL, Jansen JA and Mikos AG (2006) Modulation of cell differentiation in bone tissue engineering constructs cultured in a bioreactor, Adv. Exp. Med. Biol., 585, 225-241.

32. Bártolo PJ, Chua CK, Almeida HA, Chou SM and Lim ASC (2009) Biomanufacturing for tissue engineering: present and future trends, Virtual and Physical Prototyping, 4, 203-216.

33. Samuel RE, Lee CR, Ghivizzani S, Evans CH, Yannas IV, Olsen BR and Spector M (2002) Delivery os plasmid DNA to articular chondrocytes via novel collagen-glycosaminoglycan matrices, Human Gene Therapy, 13, 791-802.

34. Matsumoto T and Mooney DJ (2006) Cell instructive polymers, Adv Biochem Engin/Biotechnol, 102, 113-137.

35. Sanz-Herrera JA, Garcia-Aznar JM and Doblaré M (2009) On scaffold designing for bone regeneration: a computational multiscale approach, Acta Biomaterialia, 5, 219-229.

36. Hutmacher DW, Schantz JT, Lam CXF, Tan KC and Lim TC (2007) State of the art and future directions of scaffold-based bone engineering from a biomaterials perspective, J Tissue Eng Regen Med, 1, 245-260.

37. Anderson JM (1993) Mechanisms of inflammation and infection with implanted devices, Cardiovasc Pathol, 2, 33S-41S.

38. Anderson JM (1988) Inflammatory response to implants, Trans Am Soc, Intern Organs, 24, 101-107.

39. Anderson JM (1998) Biocompatibility of tissue-engineered implants, Frontiers in Tissue Engineering, Edited by C.W. Patrick, A.G. Mikos, L.V. McIntire, Elsevier.

40. Hedberg EL, Shih CK, Lemoine JJ, Timmer MD, Liebschner MAK, Jansen JA and Mikos AG (2005) In vitro degradation of porous poly(propylene fumarate)/poly(DL-lactic-co-glycolic acid) composite scaffolds, Biomaterials, 26, 3215-3225.

41. Gilbert TW, Stewart-Akers AM and Badylak SF (2007) A quantitative method for evaluating the degradation of biologic scaffold materials, Biomaterials, 28, 147-150.

42. Sung HJ, Meredith C, Johnson C and Galis ZS (2004) The effect of scaffold degradation rate on three-dimensional cell growth and angiogenesis, Biomaterials, 25, 5735-5742.

43. Domingos M, Chiellini F, Cometa S, Giglio ED, Grillo-Fernandes E, Bártolo PJ and Chiellini E (2010) Evaluation of in vitro degradation of PCL scaffolds fabricated via BioExtrusion. Part 1: Influence of the degradation environment, Virtual and Physical Prototyping, 5, 1-9.

44. Leon y Leon CA (1998) New perspectives in mercury porosimetry, Adv. Colloid Interface Sci, 76/77, 341-72.

45. Kuboki Y, Takita H, Kobayashi D, Tsuruga E, Inoue M and Murata M (1998) BMP-induced osteogenesis on the surface of hydroxyapatite with geometrically feasible and nonfeasible structures: topology of osteogenesis, J Biomed Mater Res, 39, 190-9.

46. Story BJ, Wagner WR, Gaisser DM, Cook SD and Rust-Dawicki AM (1998) *In vivo* performance of a modified CSTi dental implant coating, Int J Oral Maxillofac Implants, 13, 749-57.

47. Mikos AG, Sarakinos G, Lyman MD, Ingber DE, Vacanti JP and Langer R (1993) Prevascularization of porous biodegradable polymers, Biotechnol Bioeng, 42, 716-723.

48. Rouwkema J, Rivron NC and van Blitterswijk CA (2008) Vascularization in tissue engineering, Trends in Biotechnology, 26, 434-441.

49. Jones AC, Arns CH, Hutmacher DW, Milthorpe BK, Sheppard AP and Knackstedt MA (2009) The correlation of pore morphology, interconnectivity and physical properties of 3D ceramic scaffolds with bone ingrowth, Biomaterials, 30, 1440-1451.

50. Hollister SJ, Maddox RD and Taboas JM (2002) Optimal design and fabrication of scaffolds to mimic tissue properties and satisfy biological constraints, Biomaterials, 23, 4095-4103.

51. Lee J, Cuddihy MJ and Kotov NA (2008) Three-Dimensional Cell Culture Matrices: State of the Art, Tissue Engineering Part B, 14, 61-86.

52. Leong KF, Chua CK, Sudarmadji N and Yeong WY (2008) Engineering functionally graded tissue engineering scaffolds, Journal of The Mechanical Behavior of Biomedical Materials, 1, 140-152.

53. Oh SH, Park IK, Kim JM and Lee JH (2007) *In vitro* and *in vivo* characteristics of PCL scaffolds with pore size gradient fabricated by a centrifugation method, Biomaterials, 28, 1664-1671.

54. Yang SF, Leong KF, Du ZH and Chua CK (2001) The design of scaffolds for use in tissue engineering. Part 1, traditional factors, Tissue Engineering, 7, 679-689.

55. Wang H, Pieper J, Péters F, Blitterswijk CA and Lamme EN (2005) Synthetic scaffold morphology controls human dermal connective tissue formation, Journal of Biomedical Materials Research Part A, 74, 523-532.

56. Lawrence BJ and Madihally SV (2008) Cell colonization in 3D degradable porous matrices, Cell Adhesion & Migration, 2, 9-16.

57. Shihong LI, Wijn JRD, Jiaping LI, Layrolle P and Groot KD (2003) Macroporous biphasic calcium phosphate scaffold with high permeability/porosity ratio, Tissue Engineering, 9, 535-548.

58. O'Brien FG, Harley BA, Waller MA, Yannas IV, Gibson LJ and Prendergast PJ (2007) The effect of pore size on permeability and cell attachment in collagen scaffolds for tissue engineering, Technology and Helth Care, 15, 3-17.

59. O'Brien FG, Harley BA, Yannas IV and Gibson LJ (2005) Effect of pore size on cell adhesion in collagen-gag scaffolds, Biomaterials, 26, 433-441.

60. Stevens MM (2005) Exploring and engineering the cell surface interface, Science, 310, 1135-138.

61. Price RL, Ellison K, Haberstroh KM and Webster TJ (2004) Nanometer surface roughness increases select osteoblast adhesion on carbon nanofiber compacts, J Biomed Mater Res A, 70, 129-138.

62. Curtis ASG, Gadegaard N, Dalby MJ, Riehle MO, Wilkinson CDW and Aitchison G (2004) Cells React to Nanoscale Order and Symmetry in Their Surroundings, IEEE Trans Nanobioscience, 3, 61-65.

63. Boyan BD, Hummert TW, Dean DD and Schwartz Z (1996) Role of material surfaces in regulating bone and cartilage cell response, Biomaterials, 17, 137-146.

64. McClary KB, Ugarova T and Grainger DW (2000) Modulating fibroblast adhesion, spreading, and proliferation using self-assembled monolayer films of alkylthiolates on gold, J Biomed Mater Res, 50, 428-439.

65. Quirk RA, Chan WC, Davies MC, Tendler SJ and Shakesheff MK (2001) Poly(L-lysine)-GRGDS as a biomimetic surface modifier for poly(lactic acid), Biomaterials, 22, 865-872.

66. Chu PK, Chen JY, Wang LP and Huang N (2002) Plasma-surface modification of biomaterials, Mater Sci Eng R, 36, 143-206.
67. Yildirim ED, Besunder R, Pappas D, Allen F, Guceri S and Sun W (2010) Accelerated differentiation of osteoblast cells on polycaprolactone scaffolds driven by a combined effect of protein coating and plasma modification, Biofabrication, 2 (1).
68. Rejeb SB, Tatoulian M, Khonsari FA, Durand FA, Martel A, Lawrence JF (1998) Functionalization of nitrocellulose membranes using ammonia plasma for the covalent attachment of antibodies for use in membrane-based immunoassays, Anal Chim Acta, 376, 133-138.
69. Puleo DA, Kissling RA and Sheu MS (2002) A technique to immobilize bioactive proteins, including bone morphogenetic protein-4 (BMP-4), on titanium alloy, Biomaterials, 23, 2079-2087.
70. Daw R, O'Leary T, Kelly J, Short RD, Cambray-Deakin M, Devlin AJ, Brook IM, Scutt A and Kothari S (1999) Molecular Engineering of Surfaces by Plasma Copolymerization and Enhanced Cell Attachment and Spreading, Plasmas and Polymers, 4, 113-132.
71. Hsiue GH, Lee SD, Wang CC, Shiue MHI and Chang PCT (1993) ppHEMA-modified silicone rubber film towards improving rabbit corneal epithelial cell attachment and growth, Biomaterials, 14, 591-597.
72. Sipheia R, Martucci G, Barbarosie M and Wu C (1993) Enhanced Attachment and Growth of Human Endothelial Cells Derived from Umbilical Veins on Ammonia Plasma Modified Surfaces of Ptfe and EPTFE Synthetic Vascular Graft Biomaterials, Biomater, Artif Cell Im, 21 (4), 455.
73. Siow KS, Brichter L, Kumar S and Griesser HJ (2006) Plasma Methods for the Generation of Chemically Reactive Surfaces for Biomolecule Immobilization and Cell Colonization, Plasma Process Polym, 3, 392-418.
74. Griesser HJ, Chatelier RC, Gengenbach TR, Johnson G and Steele JG (1994) Growth of human cells on plasma polymers: Putative role of amine and amide groups, J Biomater Sci Polym, 5, 531-554.
75. Lopez LC, Gristina R, Ceccone G, Rossi F, Favia P and d'Agostino R (2005) Immobilization of RGD peptides on stable plasma-deposited acrylic acid coatings for biomedical devices, Surf Coat Technol, 200, 1000-1004.
76. De Bartolo L, Morelli S, Lopez LC, Giorno L, Campana C, Salerno S, Rende M, Favia P, Detomaso L, Gristina R, d'Agostino R and Drioli E (2005) Biotransformation and liver-specific functions of human hepatocytes in culture on RGD-immobilized plasma-processed membranes, Biomaterials, 26, 4432-4441.
77. Sokolsky-Papkov M, Agashi K, Olaye A, Shakesheff K and Domb AJ (2007) Polymer carriers for drug delivery in tissue engineering, Advanced Drug Delivery Reviews, 59, 187-206.
78. Lode A, Wolf-Brandstetter C, Reinstorf A, Bernhardt A, Konig U, Pompe W, Gelinsky M (2007) Calcium phosphate bone cements, functionalized with VEGF: Release kinetics and biological activity, J Biomed Mater Res A, 81, 474-483.
79. Murphy WL, Peters MC, Kohn DH and Mooney DJ (2000) Sustained release of vascular endothelial growth factor from mineralized poly(lactide-co-glycolide) scaffolds for tissue engineering, Biomaterials, 21, 2521-2527.
80. Kanczler JM, Barry J, Ginty P, Howdle SM, Shakesheff KM, Oreffo ROC (2007) Supercritical carbon dioxide generated vascular endothelial growth factor encapsulated poly(DL-lactic acid) scaffods induce angiogenesis in vitro, Biochem Biophys Res Commun, 352, 135-141.
81. Gu F (2004) Sustained delivery of vascular endothelial growth factor with alginate beads, J Control Release, 96, 463-472.
82. Arm DM, Tencer AF, Bain SD and Celino D (1996) Effect of controlled release of platelet-derived growth factor from a porous hydroxyapatite implant on bone ingrowth, Biomaterials, 17, 703-709.

83. Delgado JJ, Evora C, Sanchez E, Baro M and Delgado A (2006) Validation of a method for non-invasive *in vivo* measurement of growth factor release from a local delivery system in bone, J Control Release, 114, 223-229.

84. Nakahara T, Nakamura T, Kobayashi E, Inoue M, Shigeno K, Tabata Y, Eto K and Shimizu Y (2003) Novel approach to regeneration of periodontal tissues based on in situ tissue engineering: effects of controlled release of basic fibroblast growth factor from a sandwich membrane, Tissue Engineering, 9, 153-162.

85. Wei GB, Jin QM, Giannobile WV and Ma PX (2006) Nano-fibrous scaffold for controlled delivery of recombinant human PDGF-BB, J Control Release, 112, 103-110.

86. Mogi M, Kondo A, Kinpara K and Togari A (2000) Antiapoptotic action of nerve growth factor in mouse osteoblastic cell line, Life Sci, 67, 1197-1206.

87. Letic-Gavrilovic A, Piattelli A and Abe K (2003) Nerve growth factor beta(NGF beta) delivery via a collagen/hydroxyapatite (Col/HAp) composite and its effects on new bone ingrowth, J Mater Sci Mater Med, 14, 95-102.

88. Begley DJ (2004) Delivery of therapeutic agents to the central nervous system: the problems and the possibilities, Pharmacol Ther, 104, 29-45.

89. Premaraj S, Mundy B, Parker-Barnes J, Winnard PL and Moursi AM (2005) Collagen gel delivery of Tgfbeta3 non-viral plasmid DNA in rat osteoblast and calvarial culture, Orthod Craniofac Res, 8, 320-322.

90. Gombotz WR, Pankey SC, Bouchard LS, Ranchalis J and Puolakkainen P (1993) Controlled release of TGF-beta 1 from a biodegradable matrix for bone regeneration, J Biomater Sci Polym, 5, 49-63.

91. Jaklenec A, Hinckfuss A, Bilgen B, Ciombor DM, Aaron R and Mathiowitz E (2008) Sequential release of bioactive IGF-I and TGF-b1 from PLGA microsphere-based scaffolds, Biomaterials, 29, 1518-1525.

92. Park H, Temenoff JS, Holland TA, Tabata Y and Mikos AG (2005) Delivery of TGF b1 and chondrocytes via injectable, biodegradable hydrogels for cartilage tissue engineering applications, Biomaterials, 26, 7095-7103.

93. Li C, Vepari C, Jin HJ, Kim HJ and Kaplan DL (2006) Electrospun silk-BMP-2 scaffolds for bone tissue engineering, Biomaterials, 27, 3115-3124.

94. Yilgor P, Tuzlakoglu K, Reis RL, Hasirci N and Hasirci V (2009) Incorporation of a sequential BMP-2/BMP-7 delivery system into chitosan-based scaffolds for bone tissue engineering, Biomaterials, 30, 3551-3559.

95. Chen B, Lin H, Wang J, Zhao Y, Wang B, Zhao W, Sun W and Dai J (2007) Homogeneous osteogenesis and bone regeneration by demineralized bone matrix loading with collagen-targeting bone morphogenetic protein-2, Biomaterials, 28, 1027-1035.

96. Rai B, Teoh SH, Ho KH, Hutmacher DW, Cao T, Chen F, Yacob K (2004) The effect of rhBMP-2 on canine osteoblasts seeded onto 3D bioactive polycaprolactone scaffolds, Biomaterials, 25, 5499-5506.

97. Shen H, Hu X, Bei J and Wang S (2008) The immobilization of basic fibroblast growth factor on plasmatreated poly(lactide-co-glycolide), Biomaterials, 29, 2388-2399.

98. Delong SA, Moon JJ and West JL (2005) Covalently immobilized gradients of bFGF on hydrogel scaffolds for directed cell migration, Biomaterials, 26, 3227-3234.

99. Leong KF, Cheah CM and Chua CK (2003) Solid freeform fabrication of the three-dimensional scaffolds for engineering replacement tissues and organs, Biomaterials, 24, 2363-2378.

100. Bártolo PJ, Almeida HA and Laoui T (2009) Rapid prototyping and manufacturing for tissue engineering scaffolds, Int J Computer Applications in Technology, 36, 1-9.

101. Wiria FE, Chua CK, Leong KF, Quah ZY, Chandrasekaran M, Lee MW (2008) Improved biocomposite development of poly(vinyl alcohol) and hydroxyapatite for tissue engineering scaffold fabrication using selective laser sintering, J Mater Sci: Mater Med, 19, 989-996.

102. Nair LS and Laurencin CT (2006) Polymers as biomaterials for tissue engineering and controlled drug delivery, Adv Biochem Engin/Biotechnol, 102, 47-90.

103. Velema J and Kaplan D (2006) Biopolymer-based biomaterials as scaffolds for tissue engineering, Adv Biochem Engin/Biotechnol, 102, 187-238.
104. Chan G and Mooney DJ (2008) New materials for tissue engineering: towards greater control over the biological response, Trends in Biotechnology, 26, 382-392.
105. Bártolo PJ (2001) Optical approaches to macroscopic and microscopic engineering, PhD Thesis, University of Reading, UK.
106. Bártolo PJ, Mendes A and Jardini A (2004) Bio-prototyping, Design and Nature II - Comparing design in nature with science and engineering, Edited by CA Brebbia, L Sucharov, P Pascolo, WIT Press.
107. Ritman EL (2004) Micro-computed tomography - Current status and developments, Annual Review of Biomedical Engineering, 6, 185-208.
108. Potter HG, Nestor BJ, Sofka CM, Ho ST, Peters LE, Salvati EA (2004) Magnetic Resonance Imaging After Total Hip Arthroplasty: Evaluation of Periprosthetic Soft Tissue, The Journal of Bone and Joint Surgery, 86, 1947-1954.
109. Fenster A, Downey DB (2002) 3-D ultrasound imaging: a review, Engineering in Medicine and Biology Magazine, 15 (6), 41-51.
110. McElroy DP, MacDonald LR, Beekman FJ, Yuchuan W, Patt BE, Iwanczyk JS, Tsui BMW, and Hoffman EJ (2002) Performance evaluation of A-SPECT: a high resolution desktop pinhole SPECT system for imaging small animals, Nuclear Science, 49 (5), 2139-2147.
111. Edinger M, Cao Y, Hornig YS, Jenkins DE, Verneris MR, Bachmann MH, Negrin RS, Contag CH (2002) Advancing animal models of neoplasia through *in vivo* bioluminescence imaging, European Journal of Cancer, 38 (16), 2128-2136.
112. Chua CK, Leong KF and Lim CS (2003) Rapid prototyping: principles and applications, World Science Publishing, Singapore.
113. Alves NM and Bártolo PJ (2006) Integrated computational tools for virtual and physical automatic construction, Automation in Construction, 15, 257-271.
114. Szilvási-Nagy M and Mátyási G (2003) Analysis of STL files, Mathematical and Computer Modelling 38, 945-960.
115. Zhang LC, Han M and Huang SH (2003) CS file - an improvement interface between CAD and rapid prototyping systems, Int J Adv Manuf Technol, 21, 15-19.
116. Chen YH (1999) Y Z Data reduction in integrated reverse engineering and rapid prototyping, Int Journal of Computer Integrated Manufacturing, 12, 97-103.
117. MK Agoston (1976) Algebraic Topology, Marcel Dekker, New York.
118. Jackson TR, Liu H, Patrikalakis NM (1999) EM Sachs and MJ Cima, Modeling and designing functionally graded material components for fabrication with local composition control, Materials in Design, 20, 63-75.
119. Zhou MY, Xi JT and Yan JQ (2004) Modeling and processing of functionally graded materials for rapid prototyping, Journal of Materials Processing Technology, 146, 396-402.
120. Wu XJ, Liu WJ and Wang MY (2007) Modeling heterogeneous objects in CAD, Computer-Aided Design & Applications, 4, 731-740.
121. Cai S and Xi J (2009) Morphology-controllable modeling approach for porous scaffold structure in tissue engineering, Virtual and Physical Prototyping, 4, 149-163.
122. He J, Li D, Liu Y, Gong H, Lu B (2008) Indirect fabrication of microstructured chitosan-gelatin scaffolds using rapid prototyping, Virtual and Physical Prototyping, 3, 159-166.
123. Bártolo PJ (2006) State of the art of solid freeform fabrication for soft and hard tissue engineering, Design and Nature III: Comparing Design in Nature with Science and Engineering, WIT Press, UK.
124. Bártolo PJ and Mitchell G (2003) Stereo-thermal-lithography, Rapid Prototyping Journal, 9,150-156.
125. Deshmukh S and Gandhi PS (2009) Optomechanical scanning system for micro-stereolithography (MSL): analysis and experimental verification, Journal of Materials Processing Technology, 209, 1275-1285.

126. Kang H-W, Seol Y-J, Cho D-W (2009) Development of an indirect solid freeform fabrication process based on microstereolithography for 3D porous scaffolds, J Micromech Microeng, 19 (1), doi: 10.1088/0960-1317/19/1/015011.
127. Levy RA, Chu TG, Holloran JW (1997) SE Feinberg and S Hollister, CT-generated porous hydroxyapatite orbital floor prosthesis as a prototype bioimplant, American Journal of Neuroradiology, 18, 1522-1525.
128. Griffith ML and Halloran JW (1996) Freeform fabrication of ceramics via stereolithography, Journal of the American Ceramic Society, 79, 2601-2608.
129. Chu TG, Halloran JW, Hollister SJ, and Feinberg SE (2001) Hydroxyapatite implants with designed internal architecture, Journal of Materials Science: Materials in Medicine, 12, 471-478.
130. Cooke MN, Fisher JP, Dean D, Rimnac C and Mikos AG (2002) Use of stereolithography to manufacture critical-sized 3D biodegradable scaffolds for bone ingrowth, Journal of Biomedical Materials Research Part B: Applied Biomaterials, 64B, 65-69.
131. Lan PX, Lee JW, Seol YJ, Cho DW (2009) Development of 3D PPF/DEF scaffolds using micro-stereolithography and surface modification, J Mater Sci: Mater Med, 20, 271-279.
132. Melchels FP, Feijen J and Grijpma DW (2009) A poly(D,L-lactide) resin for the preparation of tissue engineering scaffolds by stereolithography, Biomaterials, 30, 3801-3809.
133. Liu VA and Bhatia SN (2002) Three-dimensional patterning of hydrogels containing living cells, Biomed Microdevices, 4, 257-266.
134. Bartolo PJ (2008) Multimaterial microstereo-termo-litografia (microSTLG), Research project financed by the Portuguese Foundation for Science and Technology (FCT).
135. Williams JM, Adewunmi A, Schek RM, Flanagan CL, Krebsbach PH, Feinberg SE, Hollister SJ and Das S (2005) Bone tissue engineering using polycaprolactone scaffolds fabricated via selective laser sintering, Biomaterials, 26, 4817-4827.
136. Lee G and Barlow JW (1996) Selective laser sintering of bioceramic materials for implants, Proceedings of the '96 SFF Symposium, Austin.
137. Zhou WY, Lee SH, Wang M, Cheung WL and Ip WY (2008)Selective laser sintering of porous tissue engineering scaffolds from poly(L-lactide)/carbonated hydroxyapatite nanocomposite microspheres, J Mater Sci:Mater Med, 19, 2535-2540.
138. Hao L, Savalani MM, Zhang Y, Tanner KE, Harris RA (2006) Selective Laser Sintering of Hydroxyapatite Reinforced Polyethylene Composites for Bioactive Implants and Tissue Scaffold Development, Proceedings of the Institution of Mechanical Engineers, Part H: Journal of Engineering in Medicine, 220 (4), 521-531.
139. Naing MW, Chua CK and Leong KF (2008) Computer aided tissue engineering scaffold fabrication, Virtual Prototyping & Biomanufacturing in Medical Applications, Edited by B Bidanda and PJ Bártolo, Springer.
140. Tan KH, Chua CK, Leong KF, Cheah CM, Cheang P, Abu Bakar MS and Cha SW (2003) Scaffold development using selective laser sintering of polyetheretherketone-hydroxyapatite biocomposite blends, Biomaterials, 24, 3115-3123.
141. Crump SS (1989) Apparatus method for creating three-dimensional objects, US Pat. 5121329.
142. Hutmacher DW, Schantz T, Zein I, Ng KW, Teoh SH and Tan KC (2001)Mechanical properties and cell culture response of polycaprolactone scaffolds designed and fabricated via fused deposition modelling, Journal of Biomedical Materials Research, 55, 203-216.
143. Zein I, Hutmacher DW, Tan KC and Teoh SH (2002) Fused deposition modelling of novel scaffold architectures for tissue engineering applications, Biomaterials, 23, 1169-1185.
144. Woodfield TB, Malda J, de Wijn J, Péters F, Riesle J and van Blitterswijk CA (2004) Design of porous scaffolds for cartilage tissue engineering using a three-dimensional fiber-deposition technique, Biomaterials, 25, 4149-4161.
145. Wang F, Shor L, Darling A, Khalil S, Güçeri S and Lau A (2004) Precision deposition and characterization of cellular poly-ε-caprolactone tissue scaffolds, Rapid Prototyping Journal, 10, 42-49.

146. Yildirim ED, Besunder R, Guceri S, Allen F and Sun W (2008) Fabrication and plasma treatment of 3D polycaprolactone tissue scaffolds for enhanced cellular function, Virtual and Physical Prototyping, 3, 199-207.

147. Park S, Kim G, Jeon YC, Koh Y and Kim W (2009) 3D polycaprolactone scaffolds with controlled pore structure using a rapid prototyping system, J Mater Sci: Mater Med, 20, 229-234.

148. Rath SN, Cohn D and Hutmacher DW (2008) Comparison of chondrogenesis in static and dynamic environments using a SFF designed and fabricated PCL-PEO scaffold, Virtual and Physical Prototyping, 3, 209-219.

149. Xiong Z, Yan Y, Zhang R and Wang X (2005) Organism manufacturing engineering based on rapid prototyping principles, Rapid Prototyping Journal, 11, 160-166.

150. Vozzi G, Flaim C, Ahluwalia A and Bhatia S (2003) Fabrication of PLGA scaffolds using soft lithography and microsyringe deposition, Biomaterials, 24, 2533-2540.

151. Miranda P, Saiz E, Gryn K and Tomsia AP (2006) Sintering and robocasting of β-tricalcium phosphate scaffolds for orthopaedic applications, Acta Biomaterialia, 2, 457-466.

152. Miranda P, Pajares A, Saiz E, Tomsia AP and Guiberteau F (2008) Mechanical properties of calcium phosphate scaffolds fabricated by robocasting, J Biomed Mater Res A, 85 (1), 218 227.

153. Yan Y, Zhang R and Lin F (2003) Research and applications on bio-manufacturing, Proceedings of the 1st International Conference on Advanced Research in Virtual and Rapid Prototyping, School of Technology and Management, Leiria, Portugal.

154. C Mota, A Mateus, PJ Bártolo, H Almeida and N Ferreira, Processo e equipamento de fabrico rápido por bioextrusão/Process and equipment for rapid fabrication through bioextrusion', Portuguese Patent nº104247, 2010

155. Domingos M, Dinucci D, Cometa S, Alderighi M, Bártolo PJ and Chiellini F (2009) Polycaprolactone scaffolds fabricated via bioextrusion for tissue engineering applications, International Journal of Biomaterials, 2009, 1-9.

156. Domingos M, Chiellini F, Gloria A, Ambrosio L, Bártolo PJ and Chiellini E (2010) Bioextruder: study of the influence of process parameters on PCL scaffolds properties, Innovative Developments in Design and Manufacturing, Edited by PJ Bártolo et al, CRC Press.

157. Sachs EM, Haggerty JS, Cima MS, Williams PA (1989) Three-dimensional printing techniques, US Pat. 5204055.

158. Kim SS, Utsunomiya H, Koski JA, Wu BM, Cima MJ, Sohn J, Mukai K, Griffith LG and Vacanti JP (1998) Survival and function of hepatocytes o a novel three-dimensional synthetic biodegradable polymer scaffolds with an intrinsic network of channels, Annals of Surgery, 228, 8-13.

159. Lam CX, Mo XM, Teoh SH and Hutmacher DW (2002) Scaffold development using 3D printing with a starch-based polymer, Materials Science and Engineering, 20, 49-56.

160. Leukers B, Gülkan H, Irsen SH, Milz S, Tille C, Schieker M and Seitz H (2005) Hydroxyapatite scaffolds for bone tissue engineering made by 3D printing, Journal of Materials Science: Materials in Medicine, 16 (12), 1121-1124.

161. Cui X, Human microvasculature fabrication using thermal inkjet printing technology, Biomaterials, 30, 6221-6227, 2009

162. Sachlos E, Reis N, Ainsley C, Derby B and Czernuszka JT (2003) Novel collagen scaffolds with predefined internal morphology made by solid freeform fabrication, Biomaterials, 24 (8), 1487-1497.

163. Lee M, Dunn JC, Wu BM (2005) Scaffold fabrication by indirect three-dimensional printing, Biomaterials, 26 (20), 4281-4289.

164. Mironov V (2009) Biofabrication: a 21st Century Manufacturing Paradigm, Biofabrication, 1.

165. Williams D (2009) On the nature of biomaterials, Biomaterials, 30 (30), 5897-5909, 2009

166. Mironov V (2009) Organ printing: tissue spheroids as building blocks, Biomaterials, 30 (12), 2164-2174, 2009

167. Chisti Y (2008) Biodiesel from microalgae beats bioethanol, Trends Biotechnol, 26 (3), 126-131, 2008.
168. Keriquel V (2010) *In vivo* bioprinting for computer- and robotic-assisted medical intervention: preliminary study in mice, Biofabrication, 2 (1), 2010.
169. Cohen DL (2006) Direct freeform fabrication of seeded hydrogels in arbitrary geometries, Tissue Eng, 12 (5), 1325-1335.